Green Energy and Technology

For further volumes:
http://www.springer.com/series/8059

Otto Andersen

Unintended Consequences of Renewable Energy

Problems to be Solved

 Springer

Otto Andersen
Western Norway Research Institute
Sogndal
Norway

ISSN 1865-3529   ISSN 1865-3537 (electronic)
ISBN 978-1-4471-7004-4   ISBN 978-1-4471-5532-4 (eBook)
DOI 10.1007/978-1-4471-5532-4
Springer London Heidelberg New York Dordrecht

Springer is part of Springer Science+Business Media (www.springer.com)

Preface

It is well known that fossil-based energy has multiple, negative environmental consequences. From initial production, through to distribution and final use, fossil fuels (coal, oil, gas) are the source of emissions and spills that cause serious impacts on both the environment and our health. Renewable energy is, in contrast, commonly held to be a clean form of energy. To substantiate this claim, it is important to scrutinize the environmental impacts of all forms of renewable energy. This includes assessing the unintended consequences from initial production, through to distribution and final use. The aim of this book is to introduce the reader to the concept of unintended consequences and how this relates to renewable energy technologies. Drawing upon a series of case studies from around the world, this book also illustrates the methods and tools that can be applied to identify the unintended consequences from these technologies from development through to implementation.

Of course, people may question the urgency of the issue, why there is a need to highlight the unintended consequences of renewable energy in the first place, and the risk that such a focus will impede the development and uptake of renewable energy. A good way of framing this issue is to consider the effects of the large-scale implementation of fuels produced from biological materials, so-called biofuels. This transition has been motivated by the goal to reduce emissions of the greenhouse gas carbon dioxide (CO_2). In 2008, an international debate emerged connecting biofuel development with increases in world food prices [2]. The debate also highlighted a reduction of biological diversity, and questioned the efficiency of biofuels as a measure to reduce GHG emissions. However, such unintended consequences of the biofuels policy could have been predicted, and measures could have been taken to avoid them, much earlier. The problems connected to the large-scale implementation of biodiesel had already been outlined in the mid-1990s through the Intelligent Energy Europe (IEE), a sub-programme of the Competitiveness and Innovation Programme (CIP) of the European Commission. In the IEE-ALTENER[1] project "Biodiesel in heavy-duty vehicles—Strategic plan and vehicle fleet experiments" key issues raised that were

---

[1] ALTENER is an acronym for actions on implementation of renewable energy conducted in the programme EU CIP-IEE.

thematized included the large land requirements as well as the large greenhouse gas emissions in the life cycle of biodiesel [1]. For example, it was shown that in the "worst case" scenario, a transition to biodiesel would actually increase GHG emissions. However, it was not until a decade later that these issues were factored into the biofuel debate pointed to above, and are now central to the efficiency the requirements made today of these fuels e.g., through the sustainability criteria of the EU [3].

This book demonstrates the need for researchers, policy makers and the energy industry alike, to pay greater attention to the potential unintended consequences. By obtaining knowledge of unintended consequences we put into practice a principle of *precautionary science* - to be aware of potential negative effects before any damage is done.

References

1. Andersen O, Lundli H-E, Brendehaug E, Simonsen M (1998) Biodiesel in heavy-duty vehicles—strategic plan and vehicle fleet experiments. Final report from European Commission ALTENER-project XVII/4.1030/Z/209/96/NOR. Western Norway Research Institute, Sogndal. http://www.vestforsk.no/ filearchive/rapport-18-98.pdf. Accessed 17 Sep 2013
2. Bhat K (2008) Misplaced priorities: Ethanol promotion and its unintended consequences. Harvard Int Rev Spring 30(1): 30–33
3. European Commission (2010) Directive 2009/28/EC of 23 April 2009 on the promotion of the use of energy from renewable sources and amending and subsequently repealing directives 2001/77/EC and 2003/30/EC. European Commission

Acknowledgments

This book is dedicated to the memory of Karl Georg Høyer, who died in November 2012. Karl Georg was my mentor and influential for much of the knowledge and insight that this book is founded on.

I would like to thank all my colleagues at Stiftinga Vestlandsforsking/Western Norway Research Institute (WNRI) who have supported this undertaking, and mobilized the necessary funding to make the writing of this book possible. In particular, I would like to extend a thank to Hans Jacob Walnum, for his supportivness in supportive in discussions on rebound effects.

Acknowledgments go to the funding mechanism of European Economic Area/ Norway Grants, for the project #PL0261 "Influence of Bio-components Content in Fuel on Emission of Diesel Engines and Engine oil Deterioration—BIODEG".

Thanks also to Jan Czerwinski at the Abgasprüfstelle und Motorenlabors (AFHB) Berner Fachhochschule für Technik und Informatik, Biel, Switzerland, for the valuable discussions and insights.

I thank David van der Spoel at Department of Cell and Molecular Biology, Uppsala University, and Sergio Manzetti of Fjordforsk AS, for facilitating the molecular dynamics studies.

I am in addition grateful to Anders Andrae of Huawei Electronics for providing me with insights into recent and current developments in consequential life cycle assessments.

Thanks goes to Deborah Davies for initial editing and proofreading of the chapters and also for her excellent suggestions for improving the text throughout the book.

Finally, I thank my wife Astrid, our daughter Alma and our sons Albert and Torjus, as well as the rest of my family, for their encouragement and inspiration throughout the writing of this book.

Contents

Abbreviations

aLCA	Attributional life cycle assessment
ALTENER	EU Competitiveness and Innovation Programme Intelligent Energy Europe "actions on implementation of renewable energy"
BEI	Biological exposure index
BIODEG	EEA project: Influence of bio-components content in fuel on emissions from diesel engines and engine oil deterioration
C_2F_6	Hexafluoroethane
CARB	California Air Resources Board
CdS	Cadmium sulphide
CdTe	Cadmium telluride
CCFPP	Critical cold filling pouring point/critical cold filter plugging point
CCS	Carbon capture and sequestration/storage
CFC	Chlorofluorocarbon
CGEM	Computable general equilibrium model
CH_4	Methane
cLCA	Consequential life cycle assessment
CIP	EU Competitiveness and Innovation Programme
CME	Coco methyl ester
CO	Carbon monoxide
CO_2	Carbon dioxide
CO_{2eq}	Carbon dioxide equivalents
CSP	Concentrating solar power
EC	European Commission
EEA	European Economic Area
EIA	Environmental impact assessment
EPA	US Environmental Protection Agency
EU	European Union
F^-	Fluoride
FAME	Fatty acid methyl ester
F_2	Fluorine
GaAs	Gallium arsenide
GHG	Greenhouse gas

GTAP	Global Trade Analysis Project
GWP	Global warming potential
GWP100	Global warming potential assuming a time horizon of 100 years
H_2O_2	Hydrogen peroxide
H_2SiF_6	Fluorosilicic acid
HF	Hydrogen fluoride
HOF	Hypofluorous acid
HORO	High oleic rape seed oil
HyNor	The hydrogen road Oslo-Stavanger
IEA	International Energy Agency
IEE	Intelligent Energy Europe
ILCD	International Reference Life Cycle Data System
IPCC	Intergovernmental Panel on Climate Change
KTH	Kungliga Tekniska Högskolan/Royal Institute of Technology
LCA	Life cycle assessment/life cycle analysis
LED	Light-emitting diode
MDS	Molecular dynamics simulation
MTC	Motor Test Center
NO_x	Nitrogen oxides
N_2O	Nitrous oxide
NCAUR	National Centre for Agricultural Utilisation Research
Nd	Neodymium
NF_3	Nitrogen trifluoride
NIMBY	Not in my back yard
NMVOC	Non-methane volatile organic compound
O_3	Ozone
OF_2	Oxygen difluoride
OME	Oleic (cis-9-octadecenoic) methyl ester
OPLS/AA	All-atom optimized potentials for liquid simulations
PAH	Polycyclic aromatic hydrocarbon
Pb	Lead
Phe	Phenanthrene
PME	Palm methyl ester
PPD	Pour point depressor
PPM	Parts per million
PV	Photovoltaic
RCN	Research Council of Norway
RME	Rape seed methyl ester
SCOEL	Scientific Committee on Occupational Exposure Limit Values
SEA	Strategic environmental assessment
SF_6	Sulphur hexafluoride
SiF_4	Silicon tetrafluoride
SiF_x	Silicon fluorides

SiO_2	Silicon oxide
SME	Soy methyl ester
SoA	State of the Art
TTW	Tank-to-Wheel
UNFCCC	United Nations Framework Convention on Climate Change
VITO	Vlaamse Instelling voor Technologisch Onderzoek
VOC	Volatile organic compound
WNRI	Western Norway Research Institute
WTW	Well-to-Wheel
ZEV	Zero-emissions vehicle

Chapter 1
Introduction: What are Unintended Consequences of Renewable Energy and How Can They be Predicted?

Abstract This introductory chapter describes the concept of unintended consequences, as well as methods and approaches that can be applied to identify the unintended consequences of renewable energy. The chapter includes a brief introduction to consequential life cycle assessment (cLCA), the study of rebound effects, and assessment of environmental impacts, so-called environmental impact assessment (EIA). Also included are other forms of relevant modeling, including molecular dynamics simulations (MDS). The types of unintended consequences addressed in this book are defined with special attention being paid to their impacts on health and the environment. The main renewable technologies covered in the book are also defined.

1.1 Unintended Consequences

Merton [33] was the first to define the concept of unintended consequences. He argued that people occasionally are so eager to realize the immediate benefits of an act, that they give no consideration whatsoever to its longer-term consequences. In a similar way, consequences can be overlooked when a person's fundamental values requires her/him to pursue change. The resulting unintended consequences of the actions may change these original values over time.

The book Green Technology [44], has an article by Ozzie Zehner, in which he describes unintended consequences in relation to green technologies as:

> Distinct unanticipated consequences that can partially or fully offset intended environmental benefits [59, p. 427]

The article by Zehner, is not only a solid and concise introduction to unintended consequences but also highly relevant to this book, due to the partial overlap between green technologies and renewable energy. Zehner describes the unintended consequences as *unplanned outcomes* that occur due to the implementation of a technology, policy, or other initiative. Unintended consequences follow from

O. Andersen, *Unintended Consequences of Renewable Energy*,
Green Energy and Technology, DOI: 10.1007/978-1-4471-5532-4_1,
© Springer-Verlag London 2013

human activities but occur at a *future time* and possibly also in a *different location*. Therefore, they can be difficult to identify or directly link to a triggering activity.

The book Risk Society by Ulrich Beck [2], provides an additional key contribution to understanding the risks inherent in the application of modern technology, and as such also raises awareness of unintended consequences. Central to Beck's 'risk society' is that the benefits of a technology are reconsidered through an examination of the negative impacts of that technology. Within this *reflexive modernization*, it is questioned whether certain applications of technology ought to have been developed, given the uncertainty of their safety [34]. However, the issue at stake in Beck's book is societal risks from modern technology as a whole, he does not focus specifically on the unintended consequences of key renewable energy technologies.

Edward Tenner's book Why Things Bite Back—Technology and the Revenge of Unintended Consequences, shows us that engineers sometimes spread the notion of *the recalcitrant machine* [54]. Captain Edward Murphy Jr., who was an engineer at Edwards Air Force Base in the United States, was a devoted believer in technological improvements. His boss, Major John Paul Stapp was a biophysicist and medical doctor who tested high-deceleration stress on himself. He had just exceeded the old record of 31 times the force of gravity on his rocket sled, but nobody could tell by how much, because the gages were not operating properly. Murphy found out that a technician had installed each of them backward. He thus concluded: "If there is more than one way to do a job and one of those ways will end up in disaster, then somebody will do it that way". Stapp later referred to this as "Murphy's Law", which he expressed more succinctly as "If anything can go wrong, it will". The term passed into the technological folklore, and aircraft companies began advertising their products as "exempt from Murphy's Law". A direct implication for engineering was the redesigning of sensors so they could only be attached one way—the correct way. This "fool proofing" is a form of *precautionary design*, which according to the Oxford English Dictionary, dates back to the early years of the automotive industry, where the concept of "fool-proof" was used in a 1902 book on automotive designs [54].

Tenner states that Murphy's Law is not a fatalistic, defeatist principle, but rather a call for needed, technological improvements. Thus, it can be considered a practical reminder of the need for methodologies and practices that look to avoid the unintended consequences of technologies.

In the Special Report of the Intergovernmental Panel on Climate Change, "Renewable Energy Sources and Climate Change Mitigation" unintended consequences is treated in the context of *interactions between energy and climate policies* [25]. The report highlights the overlapping drivers and rationales for the deployment of renewable energy, as well as overlapping jurisdictions (local, national, international). These can lead to substantial interplay among policies, and in turn give rise to unintended consequences (see also [42]). Thus, the IPCC reports states that a clear understanding of the interplay among policies and the *cumulative effects* of multiple policies, is crucial. The report supports the call for a multi and interdisciplinary approach in the study of unintended consequences, in order to address the complex and diverse issues around the decarbonization of the built environment [7].

1.1.1 Methodology for Predicting Future Consequences

1.1.1.1 Consequential LCA

Studies of environmental consequences usually employ methodologies in line with the principles in the ISO 14040:2006 and 14044:2006 standards for Life Cycle Assessment (LCA).

When conducting a LCA of an energy form, this is commonly done either as an *attributional* LCA (aLCA) or a *consequential* LCA (cLCA). In an aLCA, all the environmental impacts created in the life cycle of the energy form are detailed and summarized.

In contrast, a cLCA goes further. It aims at elucidating the consequences of a shift in products, or in a technology, e.g. a shift from fossil energy to renewable energy technologies. Thus, cLCA has the potential to be used as a modeling tool for predicting the future environmental consequences of renewable energy. There are however clear limitations for its use [10] which are detailed in Chap. 3 in this book.

1.1.1.2 Rebound Effects

Rebound effects describe the consequence that some or all of the expected effects of measures to reduce energy use (in most cases) or the emission of climate cases and other pollution are offset by behavioral or systemic responses. A thorough introduction to rebound effects can be found in the book by Herring and Sorrell [23].

Rebound effects can be considered as a type of unintended consequence, from energy-efficiency improvements in society. The broader spectrum of effects of renewable energy technologies and their implementation, including nanotechnology and health are topics not usually included in the discussions of rebound effects. In Chap.2 the rebound effects are discussed in the context of unintended consequences of renewable energy.

1.1.1.3 Environmental Impact Assessment and Environmental Assessments

An environmental impact assessment (EIA), involves the assessment of possible positive or negative impacts that a proposed project may have on the environment. There are various methods for conducting an EIA, including environmental risk assessment and risk management [35].

EIA is defined by the European Commission as "a procedure that ensures that the environmental implications are taken into account before the decisions are made" [14]. In the construction of facilities for utilizing renewable energy, an EIA can be undertaken on the basis of Directive 2011/92/EU—the "Environmental Assessment Directive".

When the object of the assessment is public plans or programmes, the assessment can be done according to Directive 2001/42/EU—the "Strategic Environmental Assessment—SEA Directive". Common for both directives is that they aim to ensure that plans, programmes, and projects that are likely to have significant effects on the environment are being environmentally assessed before their approval or authorization.

The book by Thomas B. Fischer [17] deals with SEA through comparative analysis of practice in three countries: Britain, The Netherlands, and Germany. Use of SEA is widespread but far from systematic. There are advantages in adopting a systematic and comprehensive form of SEA. Fischer claims that only once this approach is fully understood and systematically applied, will the full benefits be achieved and unintended environmental impacts minimized.

EIAs of renewable energy projects have successfully been applied to hydroelectric power plants, predicting negative influence on water quality through changes in the concentration of dissolved oxygen, nutrient loads, and suspended sediments, as well as tidal encroachment aggravating bank erosion [51]. However a re-occurring criticism of EIAs is that they commonly underestimate the real impacts by, e.g., not including rebound effects [43].

1.1.1.4 Molecular Dynamics Simulations

Molecular dynamics simulation (MDS) is a modeling approach that takes into consideration the interaction between molecules as basis for predicting the behavior of compounds in defined environments. In Chap. 5 it is shown how MDS is used to predict how the practice of blending biodiesel with fossil diesel is effecting the toxicity of the engine exhaust. This is facilitated through modeling the formation of a new type of nanoparticle, from the uncombusted fraction of the biodiesel and the toxic polynuclear aromatic hydrocarbons (PAHs). MDS-modeling has shown that this new nanoparticle can be the reason for the increased mutagenicity of exhaust from the combustion of bioblended diesel compared to the exhaust from pure diesel, observed.

1.1.2 Defining the Types of Consequences Addressed in this Book

In order to obtain knowledge of methods and tools for predicting and avoiding unintended consequences of renewable energy, the unintended consequences need to be defined. Likewise, it is necessary to define the various types of renewable energy technologies this book deals with, which is the objective of Sect. 1.2.

The consequences of renewable energy outlined by the Intergovernmental Panel on Climate Change (IPCC) in their special report on renewable energy, and is a comprehensive point of reference [25].

Another good source for the definition of impacts from renewable energies is the book "Renewable Energy: Physics, Engineering, Environmental Impacts" by Sørensen [49].

1.1.2.1 Environmental and Health Consequences

The unintended consequences addressed in this book are limited to the environmental and health consequences covered in the following paragraphs. There are many other environmental consequences which are not specifically addressed in this book. These include: reduction in visual amenity; loss of access to space for competing users; noise during infrastructure construction and operation; and other specific impacts on local ecosystems.

Emissions to air

Use of renewable energy, in the form of bioenergy combustion, is a key contributor to air emissions. It has been shown that biomass combustion releases more than 200 different chemical pollutants, including 14 carcinogens and 4 co-carcinogens, into the atmosphere [40].

This is particularly crucial in developing nations where people cook with fuelwood over open fires, and globally it is estimated that approximately 4 billion people suffer from continuous exposure to smoke.

Emissions to air have both local, regional, and global impacts. Nitrogen oxides (NO_x) and particles are examples of mainly local emissions, while the greenhouse gases [carbon dioxide (CO_2), methane (CH_4), nitrous oxide (N_2O), many halogenated compounds] have global climate impacts, e.g. from the production phase of biofuels.

The main types of emissions to air, with their environmental and health consequences, are as follows:

Carbon dioxide (CO_2)

CO_2 exists naturally in the atmosphere and is necessary for the life on Earth. It is through uptake of CO_2 that crops can grow. It is the gas, second after water vapor, which is most important to maintain the natural greenhouse effect.

Up to 1750, the concentration of CO_2 in the atmosphere was relatively stable. But from 1750 to 2012 it has increased by more than 40 %. There is little doubt that the increase is due to human activity, primarily the burning of fossil fuels such as coal, oil, and gas. Deforestation has also contributed to the increased concentration of CO_2 in the atmosphere.

The increase in the concentration of CO_2 and other greenhouse gases in the atmosphere has enhanced the greenhouse effect. We now have what is called anthropogenic greenhouse effect.

An increase in global temperature will affect the weather conditions—including more droughts and floods. The resulting sea level rise is also affecting the natural ecosystems—which can cause problems related to food supply, accessibility and quality of drinking water, health and habitation.

Methane (CH_4)

Like carbon dioxide, methane is a greenhouse gas. With a GWP of 21, the radiative forcing of methane molecules is 21 times stronger than the corresponding radiative forcing of CO_2 (assuming a time horizon of 100 years, which is the standard in UNFCCC). A small amount of methane emission can therefore have a greater effect on climate than a large emission of CO_2. The main sources of methane emissions globally are livestock, rice paddies, landfills, melting tundra the production and transportation of natural gas, and the extraction of coal. There are also methane emissions from the combustion of fuels.

An enhanced greenhouse effect is not the only consequence of methane emissions. Methane also contributes to the formation of ground-level ozone (O_3), which is formed by CH_4, CO, NO_x, and NMVOC in the presence of sunlight. One of the most serious effects of increased concentrations of ground-level ozone is damage to agricultural crops. In addition, ozone degrades the surface of buildings. But the health consequences are of particular concern. Breathing ozone can trigger a variety of health problems including chest pain, coughing, throat irritation, and congestion. It can worsen bronchitis, emphysema, and asthma. Ground-level ozone can also reduce lung function and inflame the linings of the lungs. Repeated exposure may permanently scar lung tissue [11]. On hot sunny days, ozone can reach unhealthy levels. Even relatively low levels of ozone can cause health effects. People with lung disease, children, elderly, and people who are active outdoors, are particularly sensitive to ozone. Children are at greatest risk from exposure to ozone because their lungs are still developing and they are more likely to be active outdoors when ozone levels are high, which increases their exposure. Children are also more likely than adults to have asthma.

Nitrogen oxides (NOx)

Emissions of nitrogen oxides (NO + NO_2 = NO_x) lead to a number of environmental problems. NO_x emissions actually have impacts at the three different levels: global, regional, and local. Nitrogen oxides can affect the Earth's climate through chemical processes in the atmosphere, though there is considerable uncertainty with regards this effect, making it difficult to quantify it. It is at the regional and local level that the consequences of NOx emissions are more clearly understood. Regional impacts include water acidification, which can seriously affect fish stocks as well as plant and animal life. The acidification of waterways is directly connected to a reduction in, or complete loss of biodiversity. Acidification also damages vegetation and can corrode wood and rock. Once nitrogen air pollution reaches the ground through precipitation, it contributes significantly to the eutrophication of freshwater, coastal, and marine environments. On land its presence gives rise to unbalanced nutrient absorption in plants.

Acidification of soil is known to also produce synergistic effects with toxic wastes. While toxic waste has no apparent relationship to energy usage, acidification of soil is one factor which might increase the mobility of toxic chemicals and release them into the environment. This is of particular concern in connection with rising sea levels, which are bringing into play a second factor for toxic waste mobility: increased salinity [26].

NO_x emissions cause large, local pollution problems, one of the most serious of which directly affects human health. Locally high concentrations of NO_2 aggravate asthma and other respiratory and lung diseases. In addition to that, just as described for methane, nitrogen oxides contribute to the formation of ground-level ozone (O_3).

Nitrogen oxide emissions are particularly relevant in relation to the growing use and development of biodiesel, since this fuel has higher direct emissions of NO_x than the case is for fossil diesel. An increase in NO_x -emissions of 16 % in passenger cars is observed when switching from fossil diesel to biodiesel [29].

Nitrous oxide (N_2O)

Nitrous oxide (laughing gas) is a strong greenhouse gas, with a GWP of 310. It has particular relevance in biofuel production. N_2O is released from agricultural systems, mainly due to the use of artificial fertilizers. In addition, the use of fossil fuels (especially coal) in biomass conversion processes can result in N_2O emissions that strongly impact the GHG savings [25].

Carbon monoxide (CO)
Carbon monoxide has the acute health effect of reducing the capacity of blood to carry oxygen. High concentrations of CO in confined spaces can result in sudden death. Car occupants can be vulnerable to CO poisoning if the exhaust pipe is blocked and the engine is idling, for example, in a traffic jam [6]. Chronic exposure to lower concentrations may cause persistent headaches, light-headedness, depression, confusion, memory loss, nausea, drowsiness, delayed reaction time, and vomiting [16].

When oxidized to CO_2, the CO gas becomes a contributing greenhouse gas. Together with NO_x and hydrocarbons, CO also contributes to the formation of photochemical oxidants, especially ozone (in the presence of sunlight).
Non-methane volatile organic compounds (NMVOC)

NMVOC is a common term for all volatile organic compounds (VOCs) except methane. The group includes benzene, formaldehyde, cyclohexane, and many other compounds. NMVOC react with other atmospheric gases, including nitrogen oxides and form photochemical oxidants. The main type of photochemical oxidant formed is ground-level ozone (tropospheric ozone). The different NMVOC compounds have varying potentials for the formation of oxidants.

Sulfur Dioxide (SO₂)

SO_2 is primarily of regional concern as the main cause for "acid rain". In the atmosphere SO_2 reacts with water to form sulfuric acid. Sulfur can also be deposited through so-called dry deposits, which can cause damage to crops. In addition to acidification, high concentrations of SO_2 can also impact on health by worsening respiratory diseases.

Particles (particulate matter, PM)

Particulate matter is a notation for particles floating in the air for a certain time. Particles, together with nitrogen oxides, are the pollution components with most serious consequences for local air quality. It is common to use PM_{10} as an indicator of the parts of the airborne dust that has the greatest impact on health. The remaining airborne dust—with a diameter greater than 10 μm—may also have environmental consequences, but not in the form of significant impacts on health (the larger particles are not breathable). The fine fraction, with diameter <2.5 μm is of particular health concern because the smaller particles are not only breathable, but they are carried all the way down to the lower airways and lung. Particles with diameter less than 100 nm (0.1 μm) are called nanoparticles or ultrafine particles. Some of the smaller nanoparticles can cross the membranes of lung cells and damage the interior of the cells, including the DNA. Chapter 5 of this book further examines how the blending of biodiesel into fossil diesel can result in exhaust nanoparticles potentially with this capacity.

Emissions to soil and water

The fraction of the renewable energy infrastructure waste that is not recycled ends up in landfills. Many landfills, particularly in less developed regions, contribute to water contamination through leakage into water courses. Toxic substances, e.g., heavy metals, are often present in waste and pollute the water this way. Halogenated compounds pass through waste treatment facilities to a certain extent and end up in the sludge from the waste treatment facilities, which is then deposited in landfills. In the production of photovoltaic (PV) cells, the wafers are rinsed between the numerous chemical etching stages in the wet chemical process. The resulting rinse water contains toxic compounds which are only partially being removed in waste treatment facilities. The process recycling used batteries from electric vehicles is a growing source of emissions affecting land and water systems.

The creation of standing water bodies such as reservoirs for hydroelectric power plants, can lead to a toxic build-up of the heavy metal mercury(Hg), which is made mobile in the soil by bacteria and can then enter the food chain in the form of methyl-mercury [25]. Methyl-mercury is the form of mercury that caused the acute poisoning in Minamata, Japan, with more than 2000 victims of the "Minamata disease" originating from the methyl-mercury release.[1]

Resource and land use

Loss of *biodiversity* and habitats due to resource depletion and land use practices, is a serious consequence of many forms of renewable energy. In particular, the development of large-scale monoculture energy crops, at the expense of natural land areas has proven particularly detrimental for biodiversity [25]. Large systems for concentrating solar energy also require vast tracts of land and often are in conflict with the need to maintain biological diversity in the affected areas [52]. Water consumption by agriculture based energy crops can, in many instances, lead to pressures on water supply and competition for scarce water resources [28, 45]. This has been highlighted by the Committee on Economic and Environmental Impacts of Increasing Biofuels Production, at the U.S. National Research Council [38]. Massive amounts of water are also consumed in the manufacturing processes for PV solar cells, as detailed in Chap. 7 of this book.

The *albedo effect* is connected on to the reflectance of incoming sunlight to the surface of the Earth. The darker the land surface is, the more light and heat is absorbed, conversely lighter surfaces reflect more incoming sunlight. The absorbed light warms the atmosphere. It is uncertain how large this effect is for global warming, but it should nevertheless be taken into consideration when changing the reflectance of the land, for example through establishing plantations for energy crops.

1.2 Key Renewable Energy Technologies

The world's dependence on fossil fuels is widely acknowledged to be a major cause of rising levels of carbon dioxide in the atmosphere. There is thus an urgent need to develop energy sources with lower environmental impact, and attention is focussed on renewable energy sources [21]. The various renewable energy technologies and energy sources can be expected to be developed further and implemented at increasing scales in the future.

---

[1] Japan Ministry of Environment. Minamata Disease. The History and Measures. http://www.env.go.jp/en/chemi/hs/minamata2002/ch2.html. Accessed 2 May 2013.

A definition of renewable energy is helpful in this context. The Intergovernmental Panel on Climate Change applies the following definition:

> Renewable energy is any form of energy from solar, geophysical, or biological sources that is replenished by natural processes at a rate that equals or exceeds its rate of use. Renewable energy is obtained from the continuing or repetitive flows of energy occurring in the natural environment and includes low-carbon technologies such as solar energy, hydropower, wind, tide and waves, and ocean thermal energy, as well as renewable fuels such as biomass [25, p. 958].

Renewable energy is commonly divided into two main forms; (1) *mobile* and (2) *stationary* energy, depending on the final use of the energy. This chapter presents the key renewable energy technologies that are addressed in this book, along with their definitions.

1.2.1 Mobile Renewable Energy Technologies

Mobile energy is the energy used for transportation purposes, i.e., to propel motorized vehicles. Liquid and gaseous fuels made from renewable sources of energy, as well as electricity made from renewable energy, are the most common renewable energy carriers in which chemical or electrical energy is converted into the kinetic energy that propels vehicles. Directive 2009/28/EC of the European Union states that at least 20 % of the final energy consumption in transportation should originate from renewable sources by 2020 [13].

1.2.1.1 Hydrogen Gas

Most of the hydrogen used as a transport fuel is produced by reforming natural gas. Thus, it should not strictly be classified as a renewable energy technology. However, it is possible to produce hydrogen by mainly renewable energy sources, and this is being done at a few demonstration sites today, for example at the Utsira island off the coast of Norway. At this site, surplus electricity from the wind generator is converted to chemical energy and stored—in the form of hydrogen gas. When there is sufficient wind, electricity is produced by the island's turbines. This electricity is fed to power electrolysers, which produce hydrogen gas. The hydrogen gas is stored and used to produce electricity with the use of a fuel cell when the wind is not blowing. The electricity is then supplied to households.

The electrolysis of water for hydrogen production, if conducted using renewable energy sources, as in the Utsira example, meets the criteria of a renewable energy chain. Hydrogen production can also use biomass as an energy source, for example through reforming biogas.

Hydrogen gas is used as a transport fuel by two different motor technologies:

1. Combustion of hydrogen gas in a combustion engine
2. Consumption in a fuel cell to produce electricity that powers one or more electrical engines in the vehicle.

Chapter 4 presents a case study examining the unintended consequences in connection with hydrogen implementation as a transport fuel.

1.2.1.2 Biofuels

The most common biofuels for use in transportation biodiesel, ethanol, vegetable oil, and biogas. In the European Union, the consumption of biofuels is approximately 75 % biodiesel, 15 % ethanol, and the remaining 10 % biogas and vegetable oil [12]. Biodiesel is produced from various sources of oil seed crops, with rape, sunflower, soy and palm being the most common globally. The oils undergo a conversion (transesterification) to fatty acid methyl esters (FAME). 2nd generation biodiesel refers to fuel produced from sources other than plant oils. These other sources include waste animal fat and cooking oils, but often so-called synthetic biodiesel, produced with the Fischer-Tropps reaction is used synonymously with 2nd generation biodiesel.

Ethanol is produced through fermentation of sugar beet, sugar cane, or starch (corn, wheat grain). Newer technologies being developed (2nd generation) can utilize cellulose as feedstock [40]. Environmental impacts from ethanol production vary depending on feedstock type, but mainly results from the crop's need for high amounts of water, as well as petroleum-based inputs, pesticides, and artificially produced fertilizers, which pollutes groundwater and aquifers. Second-generation ethanol may lead to increasingly negative impacts on freshwater use, global warming, toxicity, and aquatic ecotoxicity [31]. Corn tilling practices result in soil run-off, siltation of streams and rivers, and lack of oxygen leading to hypoxia zones, for example in the Gulf of Mexico [53].

Biogas consists of a mixture of methane and carbon dioxide. It is commonly produced by bacteria which break down (digests) organic compounds in manure and waste sludge.

Additives are needed for biofuels to work satisfactorily, particularly in cold weather. The blending of biofuels with fossil fuels reduces the cold weather performance problems, and is common practice for compliance with policies for increased renewable energy implementation. However, the blending can also introduce new problems, such as new toxic exhaust emissions (Chap. 5).

1.2.1.3 Electricity for Propulsion of Automobiles

Uses of mobile renewable energy includes the conversion of electricity to mechanical energy in vehicles with battery operated electrical motors; electric vehicles. As was described above for hydrogen use, the electricity must be produced by renewable energy sources, in order for electric vehicles to qualify as part of a renewable energy chain.

In the category of electric vehicles it is also possible to include hybrid vehicles that use gasoline, or more recently, diesel engines in combination with electric motors, which are rapidly growing in popularity. Currently, also hydrogen fuel cells are being applied as a viable energy conversion techhnology to serve as a "range extender" in electric vehicles, improving the driving distance between charging.

1.2.2 Stationary Renewable Energy Technologies

Stationary energy technologies convert energy to supply stationary energy uses. Space heating and the use of electricity are the most common stationary energy uses that can be facilitated through the utilization of renewable energy. Another form of stationary energy use is the utilization of energy to run industrial processes. Renewable energy for stationary uses can be *centrally* produced, for example in hydroelectric power plants, wind power "parks", or integrated into buildings to serve *local* energy needs, through solar and thermal installations. The former is subject to a higher risk from serious unintended consequences, such as flooding disasters from potential terrorist attacks on hydroelectric dams. As a result, these technologies are potentially more destructive than decentralized, less complex, locally adapted technologies [39, 41, 50]. Local energy production is more compatible with the *proximity principle*, and tends to be an under-utilized technology [4].

The proximity principle states that the products, for example energy, should be produced close to the customer [1]. The principle is highly applicable in the context of energy-efficient stationary as well as mobile renewable energy production and consumption. If the proximity principle is not adhered to, energy is spilled in its distribution, and more energy carrier must be produced to supply its demand, resulting in more environmental impacts. This is furthermore discussed in connection with energy-efficiency and energy saving in Chap. 8.

1.2.2.1 Solar Energy

In addition to the direct use of solar radiation for heating purposes, solar energy is increasingly utilized either through conversion to electric energy in photovoltaic (PV) cells or to thermal energy (heat) in water. For the latter, the term concentrating solar power (CSP) is used when solar radiation is focused before being converted to heat. The heat is transferred to water or other fluids that is circulated through pipes in district heating systems or used to generate steam that drives generators for electricity production.

The PV wet chemistry production process uses toxic, high GWP, and explosive gases as well as corrosive liquids, in the etching of the solar wafers. Dry processes are currently being developed, where the emission of contaminated waste water is expected to be reduced substantially compared with the wet chemistry process. Photovoltaic solar cells containing cadmium telluride (CdTe) are of particular concern, because of their high toxicity [58]. Cadmium is a known human cardinogen, and stringent precautions are necessary, such as life cycle management, which involves manufacturers assuming product stewardship from beginning to end of product life, is advocated if it shall continue to be used in thin film PV cells [48]. Other toxic chemicals, such as cadmium sulfide (CdS) and gallium arsenide (GaAs), are also used in PV manufacture [40]. Because these chemicals are highly toxic and persistent in the environment for centuries, disposal and recycling of used PV cells might cause major problems.

Potential unintended consequences of CSP include accidental or emergency release of toxic chemicals used in the heat transfer system [40].

1.2.2.2 Wind Power

In wind power turbines electricity is generated from the movement of a rotor, which is propelled by air currents. Noise and rotor blade reflections from large-scale wind power "parks" with multiple turbine towers can disturb residents and are often unwelcome developments, when situated near residential communities. Such resistance, known as NIMBY (Not In my Back Yard), is common for major infrastructure installations, and modern wind power technology is no exemption from this [5, 24]. With its large structures, the wind turbines are very visible in the landscape. But also, when wind energy turbines are situated in remote areas, they are connected with unintended consequences. The required maintenance roads can e.g. open up access to ecologically sensitive areas and increased human activities, which cause environmental impacts [59].

Special concern has been raised about ecological impacts from the construction and operation of both land-based and offshore wind power plants. They both impact wildlife sites, can result in bird and bat collisions, and lead to habitat and ecosystem modifications [25, 30, 40, 47].

Offshore wind farms impact marine and fish stocks, with uncertainty of its magnitude due to lack of empirical data. Other impacts of wind energy include land and marine usage conflicts, including possible radar interference from the rotating magnets, proximal impacts in the form of noise and flicker, as well as property value impacts [25, 40].

1.2.2.3 Hydroelectric Power

The construction of hydroelectric power stations have major unintended environmental impacts. One aspect that is necessary to include when calculating the resulting climate gas emissions, is the microbial decomposition of dead organic materials in reservoir dams. Methane (CH_4) emissions from reservoirs might be substantial under certain conditions [3, 8, 19, 25]. Particularly at lower latitudes

Fig. 1.1 Emission of climate gases from electricity production from various renewable energy sources (g CO_{2eq}/kWh)

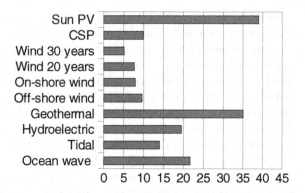

and tropical areas. These, together with N_2O emissions, can contribute substantially to the life cycle climate impacts of hydro power [20]. These aspects result is higher CO_{2eq}-emissions than several other renewable sources for electricity production, as shown in Fig. 1.1.

Hydroelectric power plants strongly affect the ecology of rivers, by inducing changes in their hydrological characteristics and by disrupting the ecological continuity of sediment transport and fish migration [25, 46, 51]. Loss of biodiversity is thus a prominent impact of hydroelectric power plants [36, 37, 57]. The operation of the power plants changes the water environment, with the consequence of reduced fish stocks due to changes in natural water flow patterns and water temperatures [22]. Fjords are particularly vulnerable to this, as exemplified with the Sognefjord in Norway [32].

1.2.2.4 Geothermal Energy

Utilization of geothermal energy resources are done in various ways depending on the geological environment, temperatures, and depths. Many high-temperature hydrothermal systems are present near areas with recent volcanic activity, i.e. near plate tectonic boundaries. Low-temperature (<100 °C) systems exist also in continental areas. Both types can be used for electricity generation, as well as direct use of the heat [55].

Sludge from geothermal drilling contains many elements of concern, incl. arsenic (As), boron (B), lead (Pb), mercury (Hg), radon (Rn), and vanadium (V), of which all are quite toxic [40].

1.2.2.5 Ocean Energy

Although not utilised extensively, ocean energy can be harvested from wave, tidal, and ocean currents, as well as salinity gradients and temperature gradients. The environmental consequences of ocean energy include noise and vibration during facility construction and decommissioning, while operational impacts might include electromagnetic fields around devices and electrical connection cables that may be problematic to sharks, skates, and rays (*elasmobranchii*) that use electromagnetic fields to navigate and locate prey [25]. Various types of water pollution, originating from the abrasion of paints and antifouling chemicals, and from chemical leakage, for example oil leaks from hydraulic systems, are other potential impacts.

1.2.2.6 Comparing Electricity Production from Various Forms of Renewable Energy

When LCA studies are compared, it is clear that the various technologies are not equally "green". In Fig. 1.1 average results form several LCA studies were compiled [15, 18, 27, 56].

As seen in Fig. 1.1, the life cycle climate gas emissions of electricity produced from PV cells and geothermal sources are high compared with various forms of wind energy. Utilization of CSP, hydroelectric, tidal, and ocean wave energy are intermediate. However, since Fig. 1.1 depicts average values, the extreme cases are not shown. They can be quite different, for example hydroelectric energy has been shown to potentially emit more than 150 g CO_{2eq}/kWh [25]. This is significantly higher than the average of the results for any of the other technologies, with PV solar as the highest (39 g CO_{2eq}/kWh). A comparable situation, although not that extreme, exists for on-shore wind energy, with highest estimates above 80 g CO_{2eq}/kWh [9, 25].

References

1. Andersen O (2003) Transport and industrial ecology: problems and prospects. PhD thesis. Vestlandsforsking, Sogndal. Aalborg University, Aalborg. http://www.vestforsk.no/en/reports/transport-and-industrial-ecology-problems-and-prospects. Accessed 17 sep 2013
2. Beck U (1992) Risk society: towards a new modernity. Sage, Newbury Park
3. Blais A, Lorrain S, Plourde Y, Varfalvy L (2005) Organic carbon densities of soils and vegetation of tropical, temperate and boreal forests. Greenhouse gas emissions—fluxes and processes. Springer, Berlin, pp 155–185
4. Bronin S (2012) Building-related renewable energy and the case of 360 State Street. Vand L Rev 65:1875
5. Cass N, Walker G (2009) Emotion and rationality: the characterisation and evaluation of opposition to renewable energy projects. Emot Space Soc 2:62–69
6. CDC (1996) Carbon monoxide poisonings associated with snow-obstructed vehicle exhaust systems—Philadelphia and New York City, Jan 1996. MMWR: Morbidity and Mortality Weekly Report. Centers for Disease Control and Prevention, pp 1–3
7. Davies M, Oreszczyn T (2012) The unintended consequences of decarbonising the built environment: A UK case study. Energy Build 46:80–85
8. Descloux S, Chanudet V, Poilvé H, Grégoire A (2010) Co-assessment of biomass and soil organic carbon stocks in a future reservoir area located in Southeast Asia. Environ Monit Assess 173(1–4):723–741

9. Dolan S, Heath G (2012) Life cycle greenhouse gas emissions of utility-scale wind power: systematic review and harmonization. J Ind Ecol 16(suppl.1):136–154
10. Ekvall T (2002) Limitations of consequential LCA. In: LCA/LCM 2002 e-conference, 20–25 May 2002. Paper presented at the InLCA/LCM 2002 E-conference, 20–25 May 2002. http://www.lcacenter.org/lca-lcm/pdf/Consequential-LCA.pdf. Accessed 05 June 2013
11. EPA (2012) Ground level ozone. Health effects. Unites States Environmental Protection Agency. http://www.epa.gov/glo/health.html. Accessed 19 June 2013
12. EurObserv'ER (2008) Biofuels barometer. SYSTÈMES SOLAIRES le journal des énergies renouvelables 185. http://www.eurobserv-er.org/pdf/baro185.pdf. Accessed 17 sep 2013
13. European Commission (2010) Directive 2009/28/EC of 23 April 2009 on the promotion of the use of energy from renewable sources and amending and subsequently repealing Directives 2001/77/EC and 2003/30/EC. European Commission
14. European Commission (2013) Environmental assessment. European Commission, Brussels. http://ec.europa.eu/environment/eia/home.htm. Accessed 29 April 2013
15. EWEA (2009) Wind energy—the facts (WindFacts). European Wind Energy Association, Brussels. http://www.wind-energy-the-facts.org/. Accessed 30 June 2013
16. Fawcett T, Moon R, Fracica P, Mebane G, Theil D, Piantadosi C (1992) Warehouse workers' headache. Carbon monoxide poisoning from propane-fueled forklifts. J Occup Med 34(1):12–15
17. Fischer TB (2002) Strategic environmental assessment in transport and land use planning. Earthscan Publications, London
18. Frischknecht R, Jungbluth N, Althaus H, Doka G, Dones R, Hellweg S, Hischier R, Humbert S, Margni M, Nemecek T, Spielmann M (2007) Implementation of life cycle impact assessment methods, final report Ecoinvent v. 2.0 No. 3. Swiss Centre for Life Cycle Inventories, Duebendorf
19. Galy-Lacaux C, Delmas R, Kouadio G, Richard S, Gosse P (1999) Long term greenhouse gas emission from a hydroelectric reservoir in tropical forest regions. Global Biogeochem Cycles 13:503–5017
20. Guerin F, Gwenaël A, Tremblay A, Delmas R (2008) Nitrous oxide emissions from tropical hydroelectric reservoirs. Geophys Res Lett 35:L06404
21. Harrison R, Hester R (2003) Sustainability and environmental impact of renewable energy sources (Issues in Environmental Science and Technology), 1st edn. Royal Society of Chemistry. http://books.google.no/books?id=8vhMtp3tXk8C&printsec=frontcover&vhl=no#v=onepage&q&f=false. Accessed 17 sep 2013
22. Helland-Hansen E, Holtedahl T, Lye O (2005) Environmental effects update. NTNU, Department of Hydraulic and Environmental Engineering, Trondheim, Norway, p 219
23. Herring H, Sorrell S (eds) (2009) Energy Efficiency and Sustainable Consumption: The Rebound Effect. Palgrave Macmillan, New York
24. Hoffman N (2011) A Don Quixote tale of modern renewable energy: counties and municipalities fight to ban commercial wind power across the United States. UMKC Law Rev 79(3):717–739
25. IPCC (2012) Renewable energy sources and climate change mitigation. Special Report of the Intergovernmental Panel on Climate Change. Cambridge University Press. http://srren.ipcc-wg3.de/report/IPCC_SRREN_Full_Report.pdf. Accessed 17 sep 2013
26. Jackson T (1991) Renewable energy. Great hope or false promise? Energy Policy Jan/Feb: 1–7
27. Jacobson M (2009) Review of solutions to global warming, air pollution, and energy security. Energy Environ Sci 2:148–173
28. Koh L, Ghazoul J (2008) Biofuels, biodiversity, and people: understanding the conflicts and finding opportunities. Biol Conserv 141(10):2450–2460
29. Kousoulidou M, Ntziachristos L, Fontaras G, Martini G, Dilara P, Samaras Z (2012) Impact of biodiesel application at various blending ratios on passenger cars of different fueling technologies. Fuel 98:88–97

30. Kunz T, Arnett E, Erickson W, Hoar A, Johnson G, Larkin R, Strickland M, Thresher R, Tuttle M (2007) Ecological impacts of wind energy development on bats: questions, research needs, and hypotheses. Front Ecol Environ 5(6):315–324
31. Liang S, Xu M, Zhang T (2012) Unintended consequences of bioethanol feedstock choice in China. Bioresour Technol 125:312–317
32. Manzetti S, Stenersen JH (2010) A critical view of the environmental condition of the Sognefjord. Mar Pollut Bull 60(12):2167–2174
33. Merton RK (1936) The unanticipated consequences of purposive social action. Am Sociol Rev 1(6):894–904
34. Morris JT (2010) Risk, language, and power: The nanotechnology environmental policy case. PhD thesis, Virginia Polytechnic Institute and State University, Falls Church, VA. http:// scholar.lib.vt.edu/theses/available/etd-10042010-225927/unrestricted/Morris_JT_D_2010_ v2.pdf. Accessed 17 sep 2013
35. Morris P, Therivel R (eds) (2009) Methods of environmental impact assessment, 2nd edn. Spon Press, London. http://www.docstoc.com/docs/71241593/30592066-Methods-of-Environmental-Impact-Assessment. Accessed 17 sep 2013
36. Nilsson C, Berggren K (2000) Alterations of riparian ecosystems caused by river regulation. Bioscience 50:783–792
37. Nilsson C, Dynesius M (1993) Ecological effects of river regulation on mammals and birds: a review. Regul Rivers: Res Manag 9(1):45–53
38. NRC (2011) Renewable fuel standard. potential eonomic and environmental effects of U.S. Biofuel Policy. National Academy of Sciences, Washington, DC. http://www.nap.edu/ catalog.php?record_id=13105. Accessed 30 June 2013
39. Perrow C (1984) Normal accidents. Living with high-risk technologies. Princeton. Princeton University Press, Princeton
40. Pimentel D (2008) Biofuels, solar and wind as renewable energy systems: benefits and risks. Springer, Dordrecht
41. Posner R (2004) Catastrophe: risk and response. Oxford University Press, New York
42. Ramasra R, Nathwani J (2013) Unintended consequential effects of overlapping carbon reduction policies. Electr J 26(1):17–26
43. Reisdorph D (2011) Rebound effects & monetizing environmental impacts. Paper presented at the Life Cycle Assessment (LCA) XI, Oct 4. Power Point. Chicago, IL
44. Robbins P, Mulvaney D, Golson J (eds) (2011) Green technology. Sage, London
45. Robertson G, Dale V, Doering S, Hamburg S, Melillo J, Wander M, Parton W, Adler P, Barney J, Cruse R, Duke C, Fearnside P, Follett R, Gibbs H, Goldemberg D, Mladenoff D, Ojima D, Palmer M, Sharpley A, Wallace L, Weathers K, Wiens J, Wilhelm W (2008) Sustainable biofuels redux. Science 322(5898):49–50
46. Shaw R, Mooney P, Hurst B (2007) Three Gorges of the Yangtze River. Chongqing to Wuhan. Odyssey Publications, Hong Kong
47. Sinclair K (2003) Avian Wind Power Research. Report, 29 April 2003. National Wind Technology Center
48. Sinha P, Kriegner C, Schew W, Kaczmar S, Traister M, Wilson D (2008) Regulatory policy governing cadmium-telluride photovoltaics: a case study contrasting life-cycle management with the precautionary principle. Energy 36:381–387
49. Sørensen B (2011) Renewable energy: physics, engineering, environmental impacts, economics & planning, 4th edn. Elsevier, Burlington
50. Sovacool B (2008) The costs of failure: a preliminary assessment of major energy accidents, 1907–2007. Energy Policy 36:1802–1820
51. Sovacool B, Bulan L (2013) They'll be dammed: the sustainability implications of the Sarawak Corridor of Renewable Energy (SCORE) in Malaysia. Sustain Sci 8:121–133
52. Stein S (2012) The environmentalist's dilemma. Policy Rev 174:49–62
53. Swenson D (2008) A review of the economic rewards and risks of ethanol production. Biofuels, solar and wind as renewable energy systems: benefits and risks. Springer, New York, pp 57–78

54. Tenner E (1997) Why things bite back: technology and the revenge of unintended consequences. http://www.citeulike.org/group/2050/author/Tenner:E. Accessed 17 sep 2013
55. Tester J, Drake E, Golay M, Driscoll M, Peters W (2005) Sustainable energy—choosing among options. MIT Press, Cambridge, 978-3527408313
56. Vestas (2005) Life cycle assessments. Report. Vestas Wind Systems A/S. http://www.vestas.com/en/about-vestas/sustainability/wind-turbines-and-the-environment/life-cycle-assessment-(lca).aspx. Accessed 17 sep 2013
57. WCD (2000) Dams and development—a new framework for decision-making. World Commission on Dams, Earthscan, London
58. Zayed J, Philippe S (2009) Acute oral and inhalation toxicities in rats with cadmium telluride. Int J Toxicol 28(4):259–265
59. Zehner O (2011) Unintended consequences of green technologies. In: Robbins P et al (eds) Green technology. Sage, London, pp 427–432

Chapter 2
Rebound Effects

Abstract Rebound effects, also known as take-back effects, refer to the behavioral or systemic responses that can be experienced after a new technology or policy measure has been implemented. Such rebound effects have been experienced from efforts to reduce energy use, climate emissions, and other pollutants as well as polluting behavior. This chapter presents a detailed account of the concept of rebound effects and presents examples of different areas where rebound effects are observed. The chapter also discusses the usefulness and limitations of the concept rebound effects in relation to improving the knowledge of the unintended consequences of renewable energy. Finally, potential rebound effects connected to nanomaterials used in new types of energy harvesting technologies are presented.

2.1 The Concept of Rebound Effects

The concept of rebound effects is mainly used for a behavioral or systemic response to the implementation of a new technology. It can also be applied to other measures to reduce energy use and emissions in relation to climate change. In principle, it can be used in reference to any natural resource or environmental problem. Within energy economics, the concept of rebound effect can be traced back to Jevons [128] who observed that the introduction of a new efficient steam engine initially decreased coal consumption, which led to a drop in the price of coal (Jevons paradox). This meant that more people could afford coal, making coal economically viable for new uses, which led to an increase in coal consumption.

The concept of rebound effects can also be traced to the late 1960s—early 1970s discourse on environmental and ecological system dynamics [5, 8, 39, 42, 43]. Here it was connected to the concept of *feedback mechanisms*, which can be both positive and negative. In this instance, rebound effects are the results of manipulations with, or isolation of parts of the system, and can lead to chains of effects emerging in other parts of the system. Thus, rebound effects can be understood as unpredictable backfires against anthropogenic encroachments in nature.

O. Andersen, *Unintended Consequences of Renewable Energy*,
Green Energy and Technology, DOI: 10.1007/978-1-4471-5532-4_2,
© Springer-Verlag London 2013

According to Maxwell et al. [38], a rebound effect can be understood through the use of the IPAT equation, originally devised by Ehrlich and Holdren [8]:

I (Impact) = P (Population) *A (Affluence) *T (Technical Efficiency)

Impact refers to total environmental impact and depends on population level, average product and service consumption per capita, and the environmental efficiency of production. The rebound effect refers to the relationship between improvements in technical efficiency (T), which lead to increased per capita consumption (A).

The principles of rebound effects are currently used by the International Panel for Climate Change (IPCC) in their reports, but only in connection with mechanisms within the natural environment, not for societal effects. While the concept of rebound effects explicitly refers to effects within society, the IPCC hardly applies the concept in this context [36].

The concept of rebound effects is most commonly used within mainstream economics theory and analysis. Other, significant contributions to its understanding, come from the de-growth economists [56]. Here rebound effects are diverse. To gain a full understanding requires an interdisciplinary approach that includes fundamental research on social and technological structures and relations. A better understanding of the concept and its implications could be achieved by including other perspectives such as industrial ecology, in addition to economist and engineering views [4, 22, 34].

Sorrell [59] claims that in an "orthodox" economical analysis, rebound effects are small, thus improvements in energy productivity make relatively small contributions to economic growth. Decoupling energy consumption from economic growth is thus considered both feasible and cheap. In contrast, an "ecological perspective" suggests that rebound effects are large, and improvements in energy productivity make an important contribution to economic growth. Decoupling is thus both difficult and expensive. The ecological economics perspective implies that capital, labor, and energy are interdependent inputs that have synergistic effects on economic output. This perspective is based on the understanding that increased availability of high-quality energy sources has provided the necessary conditions for most historical improvements in economic productivity. However, the ecological perspective highlights blind spots within orthodox economic theory. These misconceptions (blind spots) are used in the design of economic models applied to support climate policies. An ecological economics perspective questions the potential for and continued reliance upon decoupling energy consumption from economic growth.

There are basically three main areas where rebound effects are addressed. These are described in the following paragraphs.

2.1.1 Energy Efficiency Measures

Jevons paradox is often used synonymously with rebound effect in literature. It implies that energy efficiency improvements, while saving energy in the short run, may in the long-run result in higher energy use. It has basis in Jevons' argument that it is a confusion of ideas to suppose that the economical use of fuel is equivalent to diminished consumption; the contrary is rather the truth [28].

Herring et al. [20] state that Jevons paradox is an observation based on economic theory and long-term historical studies, and its magnitude is a matter of considerable dispute: if it is small (i.e., the expansion of fuel consuming activities is less than 100 % of the improvement in efficiency) then energy efficiency improvements will lead to lower energy consumption, if it is large (i.e., the expansion of fuel consuming activities is greater than 100 % of the improvement in efficiency) then energy consumption will be higher. A key problem in resolving the two positions is that it is not possible to run 'control' experiments to see whether energy use is higher or lower than if there had been no efficiency improvements. A further problem is that the rebound effect has differing impacts at all levels of the economy, from the micro-economic (the consumer) to the macro-economic (the national economy), and its magnitude at all levels of the economy has not yet been determined. Nonetheless, there is mounting evidence that at the national level it is not uncommon for total resource consumption to grow even while efficiency improves, suggesting, at least, that improvements in efficiency are not necessarily sufficient for curtailing consumption (although, once again, this does not necessarily demonstrate that resource consumption grows because of improvements in efficiency) [20].

2.1.2 Measures to Reduce the Emission of Climate Gases

When energy consumers believe their energy is derived from renewable sources, they may be less concerned about conserving it. In hybrid cars, electricity replaces part of the oil-based fuel consumption. Depending on where the electricity mixture[1] is sourced, the cars can contribute to the reduction in climate gas emissions. But if these cars are used more due to this, there is a potential rebound effect [13]. This was observed in an empirical survey in Japan by Hiroyuki Ohta and Satoshi Fujii [44]. They showed that a year after purchasing what they considered to be an 'environmentally friendly' car (e.g., a Toyota Prius with a hybrid engine), drivers who bought such cars were driving 1.6 times as far as they had done with their previous vehicle [52]. As we can see, new environmental problems

---

[1] Electricity mixture is a term used for describing the composition of the electricity, with respect to how the electricity is produced. The electricity mixture for Norway is >90 % hydroelectric, while for China it is about 20 % hydroelectric and 70 % fossil sources (mainly coal power).

can emerge as a consequence of policies for reducing climate gas emissions [9]. The biofuels policy of blending biodiesel with fossil diesel has a potential "rebound" with the emergence of new type of toxic exhaust emissions. A more detailed account of this is given in Chap. 5.

Several LCAs have revealed that the production phase of biofuels emits large amounts of climate gases [51, 55]. The rebound effect in this instance sees the implementation of the biofuel policy actually working counter to the intended effect of the policy.

2.1.3 Measures to Reduce Other Environmental Pollutants

Rebound effects can also be used in connection with measures, actions, and technologies aimed at reducing the emissions of environmental pollutants other than climate gases. One striking example is the requirement for use of catalytic converters for cars. This requirement has resulted in reductions in the emissions of several polluting compounds from individual vehicles, including, NO_x, NMVOC, and others. However, the catalytic converters are made from a range of scarce metals, such as platinum and palladium. This is a problem, because these metals are only found in very small concentrations in the Earth's crust. Ores of platinum in as low a concentration as 7 parts per million (ppm) are currently being mined. Frosch and Gallopoulos [12] pointed in 1989 to that about 20 million metric tons of ore had to be refined to produce the approx. 140 tons purified platinum metal for the USA market in that year. More than 40 % of this was used in automobiles, according to Keoleian [29]. Massive mining operations are necessary to Excavate minute of the metals needed. These activities imply the movement of large volumes of earth, and leaves behind polluted tailings and ground water [24].

2.1.4 Direct, Indirect, and Society-Wide Rebound Effects

It is common to classify rebound effects as either *direct, indirect, or society-wide*. These categories will be presented in the following paragraphs.

2.1.4.1 Direct Rebound Effects

A direct or "comfort" rebound effect occurs when improvements in energy-efficiency encourages greater use of the products and services. For example, when consumers purchases a new car, which is more fuel-efficient than the old, they might drive more [45]. This is because it becomes cheaper for them to drive. The money they save can, for example, be used on fuel for car trips they earlier made by foot, bike or public transport. However, this effect has been contested with the argument that it is not accounting for income growth over the past century [33].

In some cases, the direct rebound effects are sufficiently large to cause a "backfire" – that is they lead to an overall increase in energy consumption. This has been a reoccurring issue, first noted in the nineteenth century in relation to the steam engine and more recently with electric motors, where raised productivity and energy efficiency increase society's total energy demand [58].

The direct rebound effect was first brought to the attention of energy economists by Khazzoom [30] and has been the focus of much research on rebound effects since [15, 53]. The so-called *Khazzoom-Brookes postulate* has been described as:

> With fixed real energy price, energy efficiency gains will increase energy consumption above where it would be without the gains [58]

In his examination of direct rebound effects, Sorrell [59] distinguishes between *substitution* and *income* effects. Substitution accounts for how the increase in demand for an energy service, that becomes cheaper as a result of the increase in energy efficiency, is rooted in an allocation of income to this service. Taking substitution into consideration, the direct rebound effects in the USA manufacturing energy consumption have been estimated at 24 % [3]. Research conducted in Norway provides similar results, with a direct rebound effect of about 40 % for households and 10 % for commerce [18, 21]. For industrial activities, the size of the direct rebound depends on the possibilities for input substitution, which is higher in industries with limited substitution possibilities (e.g., metallurgic industries), than in industries with larger opportunities to adjust inputs, such as in the production of chemical and mineral products [16].

Industrial cases of rebound effects are furthermore considered in the European Commission report "Addressing the Rebound Effect" [38]. The estimates for rebound effects in the UK industrial sector are 15 % [1]. A study of 30 industry sectors in the USA showed long-term direct rebound effects of 20–60 % for energy intensive sectors, with utilities, chemicals, and agriculture having the highest effects [38]. The energy intensive sectors have the largest rebound effects, because energy costs are a significant factor in production processes. At a household level, if the increase in available income, obtained as a result of the reduced price of the energy service, is used on other energy-consuming purchases, this is an *income* effect. Sometimes a *consumption allocation* effect occurs when the efficiency savings for example in a household's expenses, are reallocated to other energy-using expenses [60].

2.1.4.2 Indirect Rebound Effects

There are numerous reasons why observed energy reduction is less than anticipated. An example, indirect rebound effect, involves money saved on reduced fuel consumption being spent on other energy-intensive goods and services, such as air-conditioners and a second (or third) car in a household.

Another indirect rebound effect results from certain energy-efficiency technologies (e.g., thermal insulation) needing considerable energy in the production phase of their life cycle. One example is that in some processes for thermal insulation production there is substantial emission of the compound 1,1,1,3,3-pentafluorobutane (HFC-365mcf) with the chemical formula $CF_3CH_2CF_2CH_3$ [40]. This compound is a very strong greenhouse gas, actually 890 times stronger than CO_2 [26].

2.1.4.3 Society-Wide Rebound Effects

The sum of the direct and indirect rebound effects from energy efficiency improvements is often called the society-wide, economy-wide, or overall rebound effect. It is commonly expressed as a percentage of the expected saved from a specific measure to improve energy efficiency. An overall rebound effect of 100 % means that the expected energy savings are entirely offset, leading to zero net savings. Economic models estimate that the overall rebound effect is quite variable, and ranges from 5 % to 60 % [13, 41].

Barker et al. [2] looked at the combination of indirect and economy-wide effects, something that is also known as the macroeconomic rebound effect. They critically reviewed the energy efficiency policies of the International Energy Agency (IEA), for period 2013–2030.[2] They estimated that the total rebound effect macroeconomic plus direct effects for final energy use over this time, will be, averaged across economic sectors, around 50 % by 2030.

It is also possible to place in the category society-wide rebound effects the so-called *market effects* or *dynamic effects* described by Gottron [14]. Decreased demand for a resource can lead to a lower resource price, making new uses economically viable. For example, residential electricity was initially used mainly for lighting, but as the price dropped many new electric devices became commonly available, thus raising domestic consumption and dependence on electricity.

2.1.5 Rebound Effect Categories Based on Cause-Commonalities

Santarius [52] departs from the common economist's view and presents four other categories of rebound effects, based on commonalities in their causes. They are:

- Financial rebound effects

  Financial rebound effects refer to an increase in energy efficiency that results in an income gain, which triggers new consumption. An *income effect* may result if fuel costs fall by 50 % when a driver switches from a 6-liter to a 3-liter car. This

---

[2] "World Energy Outlook" 2006 [25].

liberates money that can be spent on increased energy use in other areas—whether for additional journeys or for other goods and services that also consume energy.

- Material rebound effects

  A greater uptake and use of efficient technologies can be accompanied by greater energy use in their manufacture, e.g., to produce efficient building insulation products or to develop new infrastructure and markets for energy-efficient products.

- Psychological rebound effects

  The shift to energy-efficient technologies can also boost the symbolic meaning of these goods and services [48]. The effect observed in the Japanese study by Ohta and Fujii [44] referred to before, where drivers who bought an 'environmentally friendly' car ended up driving significantly more, is classified as a psychological rebound effect by Santarius [52].

- Cross-factor rebound effects

  Increasing the productivity of labor or capital can also increase the demand for energy, for example through mechanization and automation that uses energy, or if the use of energy-efficient technology also involves saving time [7, 52, 54].

2.2 Critique of the Rebound Effect

In his book *The Conundrum*, David Owen critically evaluates the way the concept rebound effect has been used [46]. He makes three crucial points, quoted below:

> It is only at the micro end of the economics spectrum that the number of mathematical variables can be kept manageable. But looking for rebound only in individual consumer goods, or in closely cropped economic snapshots, is as futile and misleading as trying to analyze the global climate with a single thermometer [46]

> Miles per gallon is the wrong way to assess environmental impact of cars. Far more reviling is to consider the productivity of driving [46]

> Promoting energy efficiency without doing anything to constrain overall energy consumption will not cause overall energy consumption to fall [46]

2.3 Rebound Effects and Renewable Energy

The outcome of any measure to reduce a certain use can be affected by a rebound effect—in the case of bioenergy, if increased production of solid, liquid, and gaseous biofuels leads to lower demand for fossil fuels, this in turn could lead to lower fossil fuel prices and increased fossil fuel consumption [50, 62, 67]. The same rebound effect applies to other renewable energy technologies that displace incumbent fossil fuel technologies [36].

In the case of bioenergy, the transition from fossil sources of energy to renewable sources implies moving down the ladder in terms of specific energy content. This is due to the fact that the energy rich carbon atoms are densely packed in fossil fuels, but a renewable source such as biodiesel is more loosely packed, and is present with significant amounts of oxygen (approx 10 % in bio-diesel). Fossil hydrocarbons thus contain more carbon than biofuels do.

When the whole energy chain is taken into consideration, this does not nec-essarily change the situation, as there is much input of fossil energy to the renewable energy chains, in particular during production (e.g., raising of energy crops) and distribution.

Moving from fossil to a renewable fuel is thus not always an energy-efficiency measure. The concept of rebound effects is mainly used in connection with energy—efficiency improvements, not the opposite. Thus, it is difficult to envision its use in relation to renewable energy, when bioenergy is the issue.

A coupling between rebound effects and renewable energy is however made by Santarius [52], who points out that the transition toward renewable energy is not possible solely by converting existing production facilities. Instead, new capacity and infrastructure will need to be developed on a large scale—entirely new markets are required. Therefore, one might speak of a "new markets" effect, resembling the *market effects* or *dynamic effects* described above. A case of these types of rebound effects can be found in connection with electric vehicles—if the electricity mixture used is favorable, the large-scale introduction of these vehicles might lead to efficiency gains per kilometer driven.

To understand the implications of the rebound effect for society at large, it is necessary to have a large system perspective. For electric cars, Santarius [52] points out that one must consider not only the life cycle consisting of the pro-duction, use and disposal of electric cars, but also the construction of the new material infrastructure made necessary by the use of the vehicles. This encom-passes, for example, industries involved in producing new engines and batteries, as well as the charging infrastructure where drivers can also swap flat batteries for newly charged ones.

Even the salaries that the operators of the new charging stations use to pay for their own energy use can produce rebound effects, e.g., if their income is higher than it used to be or if more people overall now are at work. The new markets effect encompasses all the material rebound effects that are not included in the life cycle analysis of individual products [52].

In "Energy efficiency—a critical review," Herring argues that savings from energy efficiency should be invested in "green" electricity [19]. Instead of improving energy efficiency that lowers the implicit price of energy, leading to greater use, it is argued that a more effective CO_2 reduction policy is to shift to renewable energy, subsidized through a carbon tax. It is claimed that in order to limit energy consumption, long-term policies should lead to energy sufficiency (or conservation), rather than energy efficiency.

According to Herring, the goal should be less CO_2 emissions and not less energy use. This should be attained through taxes and a policy of electricity generated from renewable energy sources, in combination with energy efficiency. Consumers who buy 'green' electricity, often in combination with a free package of energy efficiency measures, should use the cash savings from the efficiency measures to pay for the extra cost of the 'green' electricity. On a national scale, the revenue from the carbon taxes could be used to promote and subsidize carbon abatement policies such as renewable energy, or for R&D into novel carbon policies such as carbon scrubbing (removal of CO_2 from flue gases).

Stocker argues that only when a future increase in energy demand can be stopped by behavioral changes, will renewable energies fulfil their ascribed role for achieving a sustainable energy system [61]. This is the conclusion from a model of possible economic and ecological effects from the substantial increase in the use of renewable energy in Austria. The results indicate that increasing the share of renewable energy sources is an important, but insufficient step toward achieving a sustainable energy system in Austria. A substantial increase in energy efficiency needs to be complemented with a reduction of residential energy consumption.

The modeling by Stocker took into account both direct and indirect rebound effects in connection with forcing renewable energy into the Austrian market, with both increased economic activity and energy consumption as result. In the study, there is a scenario that covers the expansion of renewable energy, where a large part of the energy-saving potential could be lost due to direct rebound effects such as larger living space, inefficient heating behavior, and indirect or macroeconomic rebound effects, namely higher energy consumption due to economic growth induced by investments. With additional attitude changes, the entire potential could be exhausted.

A theoretical reasoning and clarification of the energy efficiency concept, Pérez-Lombard points to problems for measuring energy efficiency both in qualitative and quantitative terms [47]. It emphasized that the concepts of efficacy, effectiveness, intensity, performance, savings and conservation, often are used improperly, thus making energy efficiency analysis ambiguous. Promoting energy efficiency without achieving energy savings is a common problem. The same is the case for the promotion of renewable energy technology, which may increase energy consumption (negative energy savings), if the energy efficiency of the technology that uses the renewable source is lower than that of an equivalent device using energy from exhaustible (e.g., fossil) sources. A typical example being promotion of the use of biomass boilers with energy efficiencies far lower than those of gas-fired boilers.

The call for reduction in the amount of energy demanded (through energy conservation) is repeated by Pérez-Lombard, in addition to reducing the energy input (technical efficiency), so that the energy used by energy systems can be reduced, minimizing the impact on the environment. In short, energy efficiency policies are not enough, and their combination with conservation policies is essential to achieve a rational use of energy, mainly through behavioral changes.

Holm and Englund [23] elucidate that there is an apparent risk that the increased use of renewable resources will result in a loss of much of the Earth's biodiversity, unless our activities are adapted in according to the IPAT equation (described initially in this chapter). That implies achievement of demographic transitions without increase in use of natural resources, as well as reaching determined sustainable levels of anthropogenic material and energy flows both locally and globally. Such a new policy is directed toward the drivers behind the rebound effect, to find a way to sustainability.

A cultural theory framework has been applied by West et al. to develop deeper understandings of how individuals' world views can inform opinions and behavior in relation to renewable energy, and the role of rebound effects in this context [66]. The four world views of the *individualist, hierarchist, egalitarian,* and *fatalist,* are applied. Individualist discourses favor competitive markets and believe the environment is tolerant to anthropogenic impacts. Egalitarian discourses favor social equality and believe nature is fragile to anthropogenic activities. Hierarchist discourses allocate particular importance to the role of institutions and regulation in regulating human–environment relations, but believe natural systems can withstand some degree of human disturbance. Fatalist discourses believe events are determined principally by fate and so conceives nature as capricious and unmanageable. Most noteworthy, the egalitarian world view, presented in the UK study, highlights the unease about the effects of 'quick fixes' to address climate change. Switching to renewable energy encouraged individuals to consume more energy, resulting in a rebound effect.

In their study of electricity from renewable sources, Franco and Salza [11], points to the problem of energy storage for transport systems. Both electricity and hydrogen use as energy carrier in vehicles constitute efficient single subsystems for energy conversion, but when taken into consideration the total systems, including the storage of the energy, the authors emphasize that a rebound effect (backfire) resulting in no net savings, can occur.

Even though the European energy policy on bioenergy indicates that there will be more energy from renewable technologies, which are expected to be less polluting, results from many life cycle assessments show that rebound effects cause the bioenergy policy to lead to worse environmental impacts for many impact categories [6].

Another type of rebound effect from measures to reduce the emission of climate gases, are so-called "toxicological rebound effects." They have been addressed for nanomaterials used in devices for the harvesting of renewable energy [35]. Development of new types of PV solar cells are, to a large extent, motivated as a climate mitigation measure, as it in many cases constitute an alternative to producing electricity from combustion of fossil fuels. Two types of nanomaterials; fullerenes (Fig. 2.1) and carbon nanotubes (Fig. 2.2), are materials of choice for the principle components as they achieve the fast charge transfers required in organic PV solar cells.

The fullerenes and CNTs are both associated with health and environmental concerns with properties that are likely to cause unintended consequences when

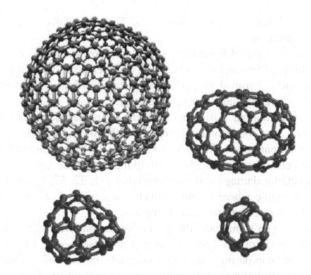

Fig. 2.1 Fullerenes, from *top left* to *bottom right*: C_{500}, C_{96}, C_{46}, C_{20} (*Source* [35])

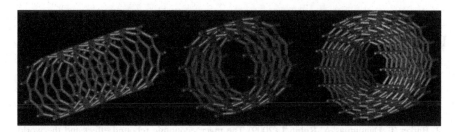

Fig. 2.2 Model of three main types of carbon nanotubes (CNTs): single walled CNT (*left*), double walled CNT (*middle*), and multi-walled CNT (*right*) (*Source* [35])

increasingly being integrated into new energy harvesting products. Fullerenes have electronic properties that give them the ability to potentially function as a vehicle for toxic compounds across blood-barriers, CNTs have unique nano-toxicological properties, resembling those of asbestos, making the compound potentially dangerous to public health and the environment during manufacture, use, and waste handling/disposal [36, 37]. Fullerenes exert damage to bacteria, plankton, cells, and multicellular organisms [10, 37, 49]. These impacts of fullerenes are primarily explained by their ability to penetrate membranes, their propensity to form aggregates, their potent reactions with different biochemical compounds and finally, by their ability to generate nanoparticles. In order to understand the nature of these unique properties, the electronic properties of fullerenes have been studied, and this has been correlated with their toxicological properties [58]. The results indicate that the chemical changes that fullerenes can undergo, in the form of changes in the cage size (interior of the fullerene molecule) are crucial for the toxicity. Such an approach to the study of fullerene toxicity can be aided with quantum chemical modeling and calculations. This is actually the case for many different nano materials.

However, the toxicity of fullerenes is not well known, and more knowledge is necessary to fully understand the correlation between toxicological properties of fullerenes and their chemical properties, size, composition, and interaction with different types of organs and tissues in the body.

CNTs are accumulating in the environment at an increasing rate, and due to their small size, they become integrated into the nutritional and reproductive environment of humans and animals [65]. CNTs, in the form of aerosol or hydrosol particles, have been shown to interfere with living organisms, and thus pose health risks through cellular damage and even mortality [31, 32, 57, 63]. It is thus clear that this is an example where rebound effects can emerge from technologies that were intended to reduce environmental impacts.

To conclude, it is a significant, though under-communicated fact, that rebound effects have not traditionally been considered in consequential life cycle assessments and when comparing the environmental impacts of various products [17, 64]. As a result, policies and practices have been put in place with counterproductive results, a fact that will be explored further in the following chapters.

References

1. 4CMR (2006) The macroeconomic rebound effect and the UK economy, report for Defra. Cambridge Centre for Climate Change Mitigation Research (4CMR), in collaboration with Cambridge Econometrics (CE) Ltd., Policy Studies Institute (PSI) and Horace Herring from the Open University
2. Barker T, Dagoumas A, Rubin J (2009) The macroeconomic rebound effect and the world economy. Energ Effi 2:411–427
3. Bentzen J (2004) Estimating the rebound effect in US manufacturing energy consumption. Energy Econ 26:123–134
4. Chertow MR (2001) The IPAT equation and its variants. J Ind Ecol 4(4). http://www.artsci.wustl.edu/~anthro/articles/jiec_4_4_13_0.pdf. Accessed 17 Sep 2013
5. Commoner B (1972) The environmental cost of economic growth. In: Population, resources, and the environment. Government Printing Office, Washington, DC, p 339–363
6. Dandres T, Gaudreault C, Tirado-Seco P, Samson R (2011) Assessing non-marginal variations with consequential LCA: application to European energy sector. Renew Sustain Energy Rev 15(6):3121–3132
7. De Fence J, Hanley N, Turner K (2009) Do productivity improvements move us along the environmental Kuznets Curve? IDEAS at the Research Division of the Federal Reserve Bank of St. Louis. http://ideas.repec.org/p/stl/stledp/2009-02.html. Accessed 17 Sep 2013
8. Ehrlich PR, Holdren JP (1971) Impact of population growth. Science 171(3977):1212–1217
9. Fölster S, Nyström J (2010) Climate policy to defeat the green paradox. Ambio 39(3):223–235
10. Fortner J, Lyon D, Sayes C, Boyd A, Falkner J, Hotze E (2005) C60 in water: nanocrystal formation and microbial response. Environ Sci Technol 39:4307–4316
11. Franco A, Salza P (2011) Strategies for optimal penetration of intermittent renewables in complex energy systems based on techno-operational objectives. Renew Energy 36:743–753
12. Frosch R, Gallopoulos N (1989) Strategies for manufacturing. Sci Am:94–102

13. Gillingham K, Kotchen M, Rapson D, Wagner G (2013) The rebound effect is overplayed. Nature 493:475–476
14. Gottron F (2001) Energy efficiency and the rebound effect: does increasing efficiency decrease demand? CRS Report for Congress. Congressional Research Service/The Library of Congress
15. Greening L, Greene D, Difiglio C (2000) Energy efficiency and consumption—the rebound effect—a survey. Energy Policy 28(6–7):389–401
16. Grepperud S, Rasmussen I (2004) A general equilibrium view of global rebound effects. Energy Econ 26(2):261–282
17. Guinee J, Heijungs R, Huppes G, Zamagni A, Masoni P, Buonamici R, Ekvall T, Rydberg T (2011) Life cycle assessment: past, present, and future. Environ Sci Technol 45(1):90–96
18. Haugland T (1996) Social benefits of financial investment support in energy conservation policy. Energy J 17(2):79–102
19. Herring H (2006) Energy efficiency—a critical view. Energy 31:10–20
20. Herring H (2006b) Jevons paradox. *Encyclopedia of Earth*. Washington, D.C: Environmental Information Coalition, National Council for Science and the Environment. http://www.eoearth.org/article/Jevons_paradox. Accessed 17 Sep 2013
21. Herring H, Roy R (2007) Technological innovation, energy efficient design and the rebound effect. Technovation 27(4):194–203
22. Hertwich EG (2005) Consumption and the rebound effect: an industrial ecology perspective. J Ind Ecol 9(1–2):85–98
23. Holm S-O, Englund G (2009) Increased ecoefficiency and gross rebound effect: evidence from USA and six European countries 1960–2002. Ecol Econ 68:879–887
24. Høyer KG (1997) Recycling: issues and possibilities. In: Brune D, Chapman D, Gwynne M, Pacyna J (eds) The global environment. Science, technology and management. VCH Verlagsgesellschaft mbH, Weinheim, pp 817–832
25. IEA (2006) World energy outook 2006. International Energy Agency, Paris
26. IPCC (2001) Third Assessment Report (TAR) of the intergovernmental panel on climate change (IPCC) prepared by Working Group III: Climate Change 2001: mitigation. Intergovernmental panel on climate change. http://www.ipcc.ch/ipccreports/tar/wg3/index.php?idp=144. Accessed 17 Sep 2013
27. IPCC (2012) Renewable energy sources and climate change mitigation. Special report of the intergovernmental panel on climate change. Cambridge University Press. http://srren.ipcc-wg3.de/report/IPCC_SRREN_Full_Report.pdf. Accessed 17 Sep 2013
28. Jevons W (1865) The coal question. Macmillan and Co, London
29. Keoleian G, Kar K, Manion M, Bulkley J (1997) Industrial ecology of the automobile. A life cycle perspective. Society of Automotive Engineers, Warrendale
30. Khazzoom J (1980) Economic implications of mandated efficiency in standards for household appliances. Energy J 1(4):21–40
31. Kwok K, Leung K, Flahaut E, Cheng J, Cheng S (2010) Chronic toxicity of double-walled carbon nanotubes to three marine organisms: influence of different dispersion methods. Nanomedicine 5(6):951–961
32. Lam C-W, James J, McCluskey R, Arepalli S, Hunter R (2006) A review of carbon nanotube toxicity and assessment of potential occupational and environmental health risks. Crit Rev Toxicol 36(3):189–217
33. Lovins A (2011) RE: The Efficiency Dilemma A letter in response to David Owen's article (December 20 & 27, 2010) (January 17.). The New Yorker. http://www.rmi.org/cms/Download.aspx?id=4952&file=2011-01_ReplyToNewYorker.pdf&title=Reply+to+%22The+Efficiency+Dilemma%22. Accessed 17 Sep 2013
34. Madlener R, Alcott B (2009) Energy rebound and economic growth:a review of the main issues and research needs. Energy 34:370–376
35. Manzetti S, Andersen O (2012) Toxicological aspects of nanomaterials used in energy harvesting consumer electronics. Renew Sustain Energy Rev 16(1):2102–2110

36. Manzetti S, Andersen O (2013) Carbon nanotubes in electronics: background and discussion for waste-handling strategies. Challenges 4(1):75–85
37. Manzetti S, Behzadi H, Andersen O, van der Spool D (2013) Fullerenes toxicity and electronic properties. Environ Chem Lett 11(2):105–118
38. Maxwell D, Owen P, McAndrew L, Muehmel K and Neubauer A (2011) Addressing the Rebound Effect. A project for the European Commission DG Environment. Final report. Ivry-sur-Seine, France: Bio Intelligence Service—Scaling sustainable development. http://ec.europa.eu/environment/eussd/pdf/rebound_effect_report.pdf. Accessed 17 Sep 2013
39. Meadows D, Meadows D, Randers J, Behrens III W (1972) The Limits to Growth. Universe Books, New York
40. Mersiowsky I, Krähling H (2002) Life cycle assessment of high-performance thermal insulation systems for domestic buildings. Paper presented at the InLCA-LCM 2002. American Center for Life Cycle Assessment. http://www.lcacenter.org/lca-lcm/session-lca.html. Accessed 17 Sep 2013
41. Nässén J, Holmberg J (2009) Quantifying the rebound effects of energy efficiency improvements and energy conserving behaviour in Sweden. Energ Effi 2:221–231
42. Odum E (1969) The strategy of ecosystem development. Science 164:262–270
43. Odum H (1971) Environment, Power, and Society. Wiley-Interscience, New York
44. Ohta H, Fujii S (2011) Does purchasing an 'Eco-car' promote increase in car-driving distance?. Unpublished Paper from the Tokyo Institute of Technology, Tokyo
45. Owen D (2010) The efficiency dilemma. If our machines use less energy, will we just use them more? (December 20.). The New Yorker. http://www.newyorker.com/reporting/2010/12/20/101220fa_fact_owen. Accessed 17 Sep 2013
46. Owen D (2012) The Conundrum. How scientific innovation, increased efficiency, and good intentions can make our energy and climate problems worse. Penguin Group, New York
47. Pérez-Lombard L, Ortiz J, Velázquez D (2013) Revisiting energy efficiency fundamentals. Energ Effi 6(2):239–254
48. Peters A, Sonnberger M, Dütschke E, Deuschle J (2012) Theoretical perspective on rebound effects from a social science point of view—Working Paper to prepare empirical psychological and sociological studies in the REBOUND project. Sraunhofer ISI. http://kooperationen.zew.de/en/rebound/results.html. Accessed 17 Sep 2013
49. Qiao R, Roberts A, Mount A, Klaine S, Ke P (2007) Translocation of C60 and its derivatives across a lipid bilayer. Nano Lett 7:614–619
50. Rajagopal D, Hochman G, Zilberman D (2011) Indirect fuel use change (IFUC) and the life cycle environmental impact of biofuel policies. Energy Policy 39:228–233
51. Reinhard J, Zah R (2009) Global environmental consequences of increased biodiesel consumption in Switzerland: consequential life cycle assessment. J Cleaner Prod 17(SUPPL. 1):46–56
52. Santarius T (2012) Green growth unraveled. How rebound effects baffle sustainability targets when the economy keeps growing. Heinrich Böll Stiftung and Wuppertal Institute for Climate, Environment and Energy, Berlin (25)
53. Saunders H (1992) The Khazzoom-Brookes postulate and neoclassical growth. Energy J 13:131–148
54. Saunders H (2000) A view from the macro side: rebound, backfire, and Khazzoom-Brookes. Energy Policy 28:439–449
55. Schmidt J (2010) Comparative life cycle assessment of rapeseed oil and palm oil. Int J Life Cycle Assess 15(2):183–197
56. Schneider F (2008) Macroscopic rebound effect as argument for economic degrowth. Paper presented at the First international conference on Economic De-growth for Ecological Sustainability and Social Equity, Paris, April 18–19th 2008. Paris. http://events.it-sudparis.eu/degrowthconference/themes/1First%20panels/Backgrounds/Schneider%20F%20Degrowth%20Paris%20april%202008%20paper.pdf. Accessed 17 Sep 2013

57. Shvedova A, Kisin E, Mercer R, Murray A, Johnson V, Potapovich A, Tyurina Y, Gorelik O, Arepalli S, Schwegler-Berry D, Hubbs A, Antonini J, Evans D, Ku B, Ramsey D, Maynard A, Kagan V, Castranova V, Baron P (2005) Unusual inflammatory and fibrogenic pulmonary responses to single-walled carbon nanotubes in mice. Am J Physiol Lung Cell Mol Physiol 289:698–708

58. Sorrell S (2007) The rebound effect: an assessment of the evidence for economy-wide energy savings from improved energy efficiency. A report produced by the Sussex Energy Group for the Technology and Policy Assessment function of the UK Energy Research Centre. UK Energy Research Centre. http://www.ukerc.ac.uk/Downloads/PDF/07/0710ReboundEffect/0710ReboundEffectReport.pdf. Accessed 17 Sep 2013

59. Sorrell S (2010) Energy, economic growth and environmental sustainability: five propositions. Sustainability 2:1784–1809

60. Steinberger J, van Niel J, Bourg D (2009) Profiting from negawatts: reducing absolute consumption and emissions through a performance-based energy economy. Energy Policy 37:361–370

61. Stocker A, Grossman A, Madlener R, Wolter M (2011) Sustainable energy development in Austria until 2020: Insights from applying the integrated model e3.at. Energy Policy 39:6082–6099

62. Stoft S (2010) Renewable fuel and the global rebound effect. Global energy policy center research paper No. 10–06. http://dx.doi.org/10.2139/ssrn.1636911. Accessed 17 Sep 2013

63. Teeguarden J, Webb-Robertson B, Waters K, Murray A, Kisin E, Varnum S, Jacobs J, Pounds J, Zanger R, Shvedova A (2011) Comparative proteomics and pulmonary toxicity of instilled single-walled carbon nanotubes, crocidolite asbestos, and ultrafine carbon black in mice. Toxicol Sci 120(1):123–135

64. Thiesen J, Christensen T, Kristensen T, Andersen R, Brunoe B, Gregersen T, Thrane M, Weidema B (2008) Rebound effects of price differences. Int J Life Cycle Assess 13(2):104–114

65. Velzeboer I, Kupryianchyk D, Peeters E, Koelmans A (2011) Community effects of carbon nanotubes in aquatic sediments. Environ Int 37:1126–1130

66. West J, Bailey I, Winter M (2010) Renewable energy policy and public perceptions of renewable energy: a cultural theory approach. Energy Policy 38:5739–5748

67. Zehner O (2011) Unintended consequences of green technologies. In: Robbins et al (eds) Green technology. Sage, London 427–432

Chapter 3
Consequential Life Cycle Environmental Impact Assessment

Abstract This chapter describes the life cycle approach to energy chain analysis and the methodology of life cycle assessment (LCA). Consequential LCA (cLCA) is discussed in comparison with attributional LCA (aLCA). The methodological approach of environmental impact assessment (EIA) is also presented. The methods, with emphasis on cLCA, are discussed in the context of improving the knowledge of unintended consequences from various forms of renewable energy. The chapter presents a series of examples where cLCA are used to predict in advance, unanticipated impacts of different forms of renewable energy technologies throughout their life cycle, with particularly focus on the impact of biofuels production.

3.1 The Life Cycle Approach to Energy Analysis

A life cycle assessment (LCA) is the compilation and evaluation of the inputs, outputs, and potential environmental impacts of a product (or energy) system throughout its life cycle [33]. LCA of an energy form thus takes into consideration the whole life cycle of the energy, from extraction/production through to refining and distribution/transmission and to its final use. These different segments in the life cycle of the energy type are also known as parts of the "energy chain". When considering the use of energy along an energy chain we are talking about two main forms of energy uses, *direct* energy use and *indirect* energy use.

Direct energy use, refers to the end use of energy. In the case of mobile energy, an example of direct energy use is diesel to power motorised vehicles. In the case of stationary energy use, an example of direct energy use is the electricity consumed to heat an electric oven. This is also known as the *net direct* energy, and is also commonly referred to as tank-to-wheel (TTW) energy.

If we move backwards along the energy chain, from distribution/transmission through to the original extraction/production of the energy, and add up the amounts of energy used for these different segments, we obtain a measure for the *gross direct* energy, or well-to-wheel (WTW) energy.

O. Andersen, *Unintended Consequences of Renewable Energy*,
Green Energy and Technology, DOI: 10.1007/978-1-4471-5532-4_3,
© Springer-Verlag London 2013

When assessing the gross direct energy use for fuels such as biodiesel, we must take into consideration the amount of energy used in the cultivation of the energy crop, as well as harvesting, pressing of seeds to obtain oil, and conversion of oil into biodiesel. Also the transport of the biodiesel from the production facility to the filling station, has to be included.

Indirect energy use refers to the energy used in the construction and maintenance of infrastructure necessary for facilitating direct energy use. This includes the production of vehicles, construction and maintenance of roads, railway lines, airports, bridges, tunnels, filling stations, and more.

When conducting a LCA of an energy type, both *direct* energy and *indirect* energy are included. This methodology for predicting the future consequences of renewable energy is guided by the principles, framework, and requirements in the ISO 14040:2006 and 14044:2006 standards for LCA. Guidelines for conducting LCA are also provided in the international reference life cycle data system (ILCD) handbook [49].

For an introduction to the LCA of energy systems, a good starting point is provided by Kummel et al. [30]. In their book, the authors discuss the methodology for performing system-wide life cycle analysis in the energy field. They applied the selected methodology to study three future energy system scenarios in Denmark, and related the results to the current energy system. Two finding relating to methodology were emphasized:

1. Conventional chain analysis used in product life cycle analysis can result in double counting, leading to erroneous results. The recommendation is to limit the analysis to the energy system components, and not include indirect effects from other parts of the economic system (e.g., agriculture delivering manure to biogas plants).
2. Two accounting principles were identified; *economy point-of-view* and *product point-of-view*. In the *economy point-of-view* all impacts from activities within the Danish economy were enumerated, but impacts from the production of materials that were imported into Denmark were omitted, and assigned as being part of the impacts from the total economy of the exporting country. In the *product point-of-view* accounting principle the impacts from imported goods were included in the Danish load, but not impacts associated with Danish export.

The study constituted the first national evaluation of the Danish energy system, and took into considerations specific Danish emission data, dispersion conditions, population densities, and the high degree of combined heating and electricity production. These aspects were not sufficiently include in previous studies, for example those of the European Commission [17, 18].

Another useful resource worth consideration is the LCA database Ecoinvent, developed by the Swiss Center for Life Cycle Inventories. The database accommodates about 4,000 datasets for products, services and processes often used in LCA case studies [1, 2, 48].

A weakness in the current use of the LCA methodology is that rebound effects (addressed in the previous chapter of the book) are not considered when comparing various products or systems [23, 45]. Inclusion of rebound effects may significantly influence the conclusions of LCAs [37, 45].

3.1.1   Attributional Versus Consequential LCA

LCAs are commonly conducted as either *attributional* LCA or as *consequential* LCA. It is worth noting that these terms are not universal. Attributional LCAs are also referred to as descriptive [24], accounting [3], or retrospective [46] LCAs, while cLCA is also know as change-oriented [24] or prospective [46] LCA.

In the aLCA of an energy system, all the environmental impacts created in the life cycle of the energy form are detailed and summarized. The focus is on describing the environmentally relevant physical flows to and from the life cycle stages and their subsystems [20]. The cLCA goes further, setting out to describe how environmentally relevant flows will change, in response to possible future decisions (e.g., energy policy implementations). The use of scenarios can improve the consistency of system choices [34, 50].

Attributional LCAs are thus limited in their capacity to highlight the environmental impacts of a product or energy system. They are, however, a convenient means to identifying and communicating improvement opportunities for existing products. On the other hand, cLCAs can highlight the indirectly induced consequences of decision making, in addition to the direct ones [5, 6].

The main characteristic of cLCA is that it provides information on the environmental consequences of a new action, or technology shift. It does this by integrating economic models to include market information. The cLCA thus represents a convergence of LCA and economic modelling approaches [12]. In contrast, the aLCA tends to look backwards at effects that have occurred, while the cLCA usually is forward-looking [38].

There is, however, an accuracy problem in that the models generated in a cLCA include aspects that are not always effects of changes [15]. For example, cLCAs are typically based on the assumption that when the demand for a material increases in the life cycle investigated, the production of that material is increased by the same amount. But materials are typically bought on a market with other suppliers and purchasers. When more material is used in the system studied, less material of that type might be used in other product systems. As a result, the increase in total production can be much smaller than a cLCA indicates [14].

In a cLCA the system is expanded to include activities both within and outside the life cycle, that are affected by changes within the life cycle of the energy form [16, 34, 35]. With regards to renewable energy technologies, cLCA can be used as a modeling tool for predicting future environmental consequences of a shift from fossil energy to renewable energy.

3.1.2 Environmental Impact Assessments

Environmental impact assessment (EIA) is another tool, in addition to LCA, for assessing the possible positive or negative impacts a proposed project or technology systems may have on the environment [20].

Unlike LCA, which is location-independent, EIA is much more a procedural tool for the evaluation of local environmental impacts [26]. An EIA generally takes into account the specific, local, geographic situation, and the existing background pressure on the environment [47]. Besides assessing quantifiable aspects, EIA also provides qualitative assessment of landscape, archaeological and cultural assets, as well as concerns of potentially affected people. It also requires involvement/participation of the public and other stakeholders in the process.

There are various methods for conducting an EIA, including the assessment and management of risks [32]. The risk aspect is of particular relevance in connection with unintended environmental consequences, as it provides knowledge of *what can go wrong*; *how likely is it*; and *what are the consequences?* [4, 28].

The European Commission defines EIA as "a procedure that ensures that the environmental implications are taken into account before the decisions are made" [19]. Its relevance to renewable energy becomes clear in the following statement:

An EIA can be undertaken for projects, such as construction of facilities for utilizing renewable energy, on the basis of Directive 2011/92/EU—the Environmental Assessment Directive [19].

When the object of the assessment is at a "higher" level of decision-making (i.e., for strategies, policies, public plans, or programs), a Strategic Environmental Assessment (SEA) can be done [26]. This is guided by the Directive 2001/42/ EU—the "Strategic Environmental Assessment–SEA Directive". Common for both directives 2011/92/EU and 2001/42/EU is that they aim to ensure plans, programs, and projects that are likely to have significant effects on the environment are environmentally assessed before their approval or authorization.

The SEA is normally conducted at an early stage, and performed in conditions involving less information and higher uncertainties. The book by Thomas B. Fischer [21] deals with SEA through comparative analysis of practice in three countries: Britain, The Netherlands, and Germany. The author observes that use of SEA is widespread, but far from systematic. Furthermore, that there are advantages to be gained from adopting a systematic application of SEA in its entirety. This is in contrast to current use, characterized by many different, and often deficient, approaches. It is claimed that only once the SEA approach is fully understood and systematically applied, will the full benefits be achieved and unintended environmental impacts minimized.

EIAs have been applied to renewable projects such as hydroelectric power plants. Here they sought to predict the negative influence on water quality, such as changes in the concentration of dissolved oxygen, nutrient loads, and suspended sediments, as well as tidal encroachment, which could aggravate bank erosion

[44]. However, EIAs have been criticised for underestimating real impacts by not example, include rebound effects [37].

The role of an impact assessment is to categorize and quantify potential environmental effects. Once this is done, deciding whether one impact is worse than another is necessarily a subjective process in which the perceptions of the decision maker are applied [8].

The IPCC states that a current weakness of EIAs for bioenergy systems, is that there is rarely a comparison to the replaced systems [25]. In addition, the methodologies and underlying assumptions for assessing environmental and socio-economic effects are not yet standardized or uniformly applied. The conclusions reached by the studies are thus inconsistent [29]. A particular challenge is that the system boundaries of the IEAs are difficult to quantify and that there are numerous interrelated factors, many of which are poorly understood or unknown [25].

3.1.3 The Use of cLCA as a Tool to Predict Unintended Consequences of Renewable Energy

3.1.3.1 Cultivating Biomass for Ethanol Production

The environmental impacts of European energy policies, including the bioenergy policy have been assessed, including the impacts of European energy generation and perturbation of world economy [9]. It was shown, that in combination with the economic general equilibrium model GTAP, cLCA could be used to predict environmental impacts, taking into account significant changes affecting large systems. For example, when considering increasing biomass production, if this land is already used for corn production, the result will be less corn production in that region. This corn must therefore be produced in a different region, and thus falls outside the scope of cLCAs because they are mostly used on small systems [10]. Therefore, a new approach, of taking into consideration large perturbations affecting large systems, such as significant substitution of European fossil fuels by renewable energies, could consist of a sequential application of computable general equilibrium model (CGEM) and cLCA [10].

Another consequence of biofuels assessed with cLCA include changes in the land use for increased cassava cultivation [43]. This is normally not included in aLCAs. Cassava, a source for ethanol production, is shown to induce land use changes that increase GHG emissions, of which about 60 % are due to changes in soil carbon stock and carbon stock loss from clearance of land. By comparing different scenarios for increased cassava production in Thailand, it was concluded that in the scenarios for expanding its cultivation area there will be significantly higher life cycle GHG emissions than in scenarios with only increased productivity of existing land for this energy crop.

3.1.3.2 Hydrogen Energy

One example where LCA has revealed some unintended consequences is in the use of hydrogen fuel in road transport. The influential WTW study by the association of European oil companies, Concawe, concluded that using hydrogen as a road transport fuel would increase Europe's greenhouse gas emissions rather than cut them, at least in the sorter time frame [13]. However, this study has been criticised for applying a mixture of attributional and consequential approach, which in addition to the short time frame has resulted in what Sanden and Karlstrom calls "somewhat misleading" interpretations [38].

3.1.3.3 The Palm Oil Controversy

It is important in the LCA results of biodiesel that the market shift in the global supply of vegetable oil, from soy, rapeseed, and sunflower, to a greater use of palm oil, is reflected. Many LCAs on rapeseed oil have been conducted, and the results are being used as decision support for bioenergy policies. Schmidt and Weidema argue, with basis in cLCA methodology, that the assessments did not, any longer reflect the real situation regarding what type of oils are actually being used [40]. Updated cLCAs could depict new knowledge of the global life cycle environmental consequences of various forms of biodiesel.

This is indeed what was experienced through a subsequent cLCA of palm and rapeseed oil, where it was shown that palm oil cultivation on peatland increases the contribution to global warming significantly, actually by a factor of 4–5 compared to cultivation on mixed soil types [39]. The other hot-spot related to global warming was treatment of the effluent from the palm oil mill.

Consequential LCA has been applied to predict the impact on GHG emissions and land use, from a replacement of either one percent of the Swiss fossil diesel use with biodiesel produced from Malaysian palm oil (palm methyl ester, PME) or from Brazilian soybean oil (soy methyl ester, SME), respectively [35]. Due to the fact that, in contrast with aLCA, cLCA uses system enlargement to include the products affected by a change of the physical flows, the environmental impacts of an increased SME consumption depend on the marginal replacement products on the world market, rather than on local production factors. The marginal products assumed to be affected are most important for the results obtained, i.e., in particular the marginal vegetable oil, fodder cake, and land areas. The authors conclude that in this perspective, it is not relevant in what country biodiesel production takes place, but rather what type of vegetable oil is involved. With respect to PME, the most relevant determining factor for the environmental impacts was shown to be the land area affected by the increased cultivation of oil palms. This expansion displaces peat land and rain forest, causing much larger life cycle GHG emissions compared with fossil diesel. In total, both PME from Malaysia and SME from Brazil were found to cause higher environmental impacts than allowed by the Swiss tax redemption on agro-biofuels (max. 60 % GHG emissions and 125 % overall environmental impact,[1] of fossil diesel).

3.1.3.4 European Production of Biodiesel

A consequential approach has been called for in the estimation of environmental impacts of biofuel production in France [22]. This is considered more relevant, to properly take into account the effects of indirect land use change and all the emissions due to biofuel production, than simply adding the effects of indirect land use change and GHG emissions to the results of an aLCA. These are conclusions after an aLCA that determined the ability of RME to reduce GHG emissions by 59 %. This result fulfils the GHG sustainable criterion of the European Directive on renewable energy (2009/28/EC). But, since the sensitivity analyses of the aLCA showed many uncertainties at the inventory analysis step, the authors predict that cLCAs most likely will contest this.

cLCA has been used to assess the environmental consequences of energetic rape use, as RME in switzerland [36]. The backdrop is that usage of rape substitutes its use as edible oil. The study showed that displacing food production by RME production can reduce total GHG emissions. However, this only occurs if GHG-intense soy meal from Brazil is substituted by rape meal, a co-product of the vegetable oil production. On the other hand, an increased production of vegetable oils was shown to increase many other environmental impact factors, because agricultural production of edible oil is associated with higher environmental impacts than the production and use of fossil fuels. The authors conclude that the environmental impacts of an increased RME production in Switzerland, depend more on the environmental scores of the marginal replacement products on the world market, than on local production factors.

3.1.3.5 Biorefineries

In a study from Finland it was shown by cLCA that the introduction of a biorefinery results in increased life cycle GHG emissions—when the production of the replacement alternatives for the raw material yields more emissions than the current system does [43]. The GHG emissions of this replacement are assumed to exceed the credits that are achieved through avoided emissions from the refining process and replacement of fossil fuels by the new biomass-based materials and fuels. By only using aLCA this effect would not have been determined.

---

[1] The term "overall environmental impact" is determined by the "environmental scarcity method", or UBP method, where UBP is an acronym for umweltbelasttungspunkt. The UBP method aggregates all environmental impacts to a single number, which represents the total environmental impact. This is an example of a so-called "EndPoint method" [7].

That second-generation biofuel are not necessarily better for the environment, is also shown in a South African study [31]. Scenarios were made for a sugar mill to start selling its bagasse, currently used to provide process heat, to an advanced biofuels producer. The sugar mill would have to buy an equivalent amount of coal to compensate for the process heat requirement. Seven scenarios, ranging from status quo, where no bagasse is diverted, to 100 % bagasse diversion, also include one scenario with energy efficiency improvements in the sugar mill. A cLCA was applied to the seven scenarios, covering GWP, non-renewable energy use, aquatic eutrophication and terrestrial acidification. A basic financial analysis of the proposed scenarios showed that they are realistic, with potentially lucrative returns. However, cLCA results showed that diverting bagasse without efficiency improvements from its current use to an ethanol biorefinery would backfire for all environmental impact categories studied. The base case outperformed all the other scenarios, with the 100 % bagasse diversion scenario emerging the worst. Thus, diverting cellulosic residues into biofuel production is not to be advised, unless accompanied by major energy efficiency improvements.

3.1.3.6 Biogas

The Consequences of changing to biogas production, and injection of the biogas into natural gas pipeline grids, has been determined by aLCA to reduce GHG emissions by 30–40 % and 10–20 %, in a 500 and 100 year time horizon, respectively [27]. However, the authors emphasize that it is necessary to conduct a cLCA to determine the consequences of changing farming activities, on the various segments of the food production chain.

3.1.3.7 Determining the Impacts of Renewable Energy in the Electricity Mix

To determine the GHG emissions of electricity production from renewable technologies is a challenge, particularly when the grid electricity comprises a complex mixture of electricity from various energy sources. The specific challenge in aLCA is to determine the appropriate electricity production mix. The selection of the data set may have significant impacts on the results.In addition, without a harmonized methodology and data management system, there is a noticeable risk of double-counting either the GHG emissions or the share of certain electricity production forms, when considering or comparing the results of various LCA studies. For example, one LCA study may use national figures, whereas another may apply figures of larger or smaller market area. In cLCA it must in addition identify the marginal technologies affected and the related consequences [42]. There are significant uncertainties involved in these assessments, particularly in future-related LCAs, where harmonization of methods and data sets are necessary, to avoid subjective choices.

References

1. Althaus H, Bauer C, Doka G, Dones R, Frischknecht R, Hellweg S, Humbert S, Jungbluth N, Köllner T, Loerincik Y, Margni M and Nemecek T (2010) Implementation of life cycle impact assessment methods. Dübendorf, CH: Swiss Centre for Life Cycle Inventories. http://www.ecoinvent.org/fileadmin/documents/en/03_LCIA-Implementation-v2.2.pdf. Accessed 02 Aug 2013
2. Althaus H, Doka G, Dones R, Hischier R, Hellweg S, Nemecek T, Rebitzer G and Spielmann M (2007) Overview and methodology. Dübendorf, CH: Swiss Centre for Life Cycle Inventories. http://www.ecoinvent.org/fileadmin/documents/en/01_OverviewAndMethodology.pdf. Accessed 17 sep 2013
3. Baumann H (1998) Life cycle assessment and decision making: theories and practices. Technical Environmental Planning. AFR report. Göteborg, Sweden: Chalmers University of Technology
4. Brookes A (2009) Environmental risk assessment and risk management (second.). In: Morris P, Therivel R (eds) Methods of environmental impact assessment. SPON PRESS, London. Taylor & Francis Group, pp 351–364. http://www.docstoc.com/docs/71241593/30592066-Methods-of-Environmental-Impact-Assessment. Accessed 17 sep 2013
5. Chen I-C, Fukushima Y, Kikuchi Y, Hirao M (2012) A graphical representation for consequential life cycle assessment of future technologies—Part 1: methodological framework. Int J Life Cycle Assess 17:119–125
6. Chen I-C, Fukushima Y, Kikuchi Y, Hirao M (2012) A graphical representation for consequential life cycle assessment of future technologies—Part 2: two case studies on choice of technologies and evaluation of technology improvements. Int J Life Cycle Assess 17:270–276
7. Climatop (2011) Climatop includes sustainability criteria. Climatop—intelligent, climatefriendly products. http://www.climatop.ch/index.php/sustainability_en.html. Accessed 30 May 2013
8. Curran M (2008) Life-cycle assessment. Human Ecology. Elsevier, pp 2168–2174
9. Dandres T, Gaudreault C, Tirado-Seco P, Samson R (2011) Assessing non-marginal variations with consequential LCA: application to European energy sector. Renew Sustain Energy Rev 15(6):3121–3132
10. Dandres T, Gaudreault C, Tirado-Seco P, Samson R (2012) Macroanalysis of the economic and environmental impacts of a 2005–2025 European Union bioenergy policy using the GTAP model and life cycle assessment. Renew Sustain Energy Rev 16(2):1180–1192
11. Dandres T (2012) Développement d'une méthode d'analyse du cycle de vie conséquentielle prospective macroscopique: évaluation d'une politique de bioénergie dans l'union européenne à l'horizon 2025. Thèse présentée en vue de l'obtention du diplôme de philosophiae doctor (génie chimique), Montreal, École Polytechnique De Montréal, Université De Montréal. http://publications.polymtl.ca/881/1/2012_ThomasDandres.pdf. Accessed 17 sep 2013
12. Earles J, Halog A (2011) Consequential life cycle assessment: a review. Int J Life Cycle Assess 16(5):114–453
13. Edwards R, Griesemann J-C, Larivé J-F and Mahieu V (2008) Well-to-wheels analysis of future automotive fuels and powertrains in the European context. CONCAWE, EUCAR and JRC
14. Ekvall T (1999) System expansion and allocation in life cycle assessment—with implications for wastepaper management. PhD Thesis, Gothenburg, Sweden, Chalmers University of Technology
15. Ekvall T (2002) Cleaner production tools: LCA and beyond. J Cleaner Prod 10:403–406
16. Ekvall T, Weidema B (2004) System boundaries and input data in consequential life cycle inventory analysis. Int J Life Cycle Assess 9(3):161–171
17. European Commission (1994) Biofuels. Report EUR 15647 EN. Brussels: DG XII

18. European Commission (1995) ExternE: Externalities of Energy. Prepared by ETSU and IER for DGXII: Science, Research & Development, Study EUR 16520-5 EN, Luxembourg
19. European Commission (2013) Environmental assessment. European Commission, Brussels. http://ec.europa.eu/environment/eia/home.htm. Accessed 29 April 2013
20. Finnveden G, Hauschild M, Ekvall T, Guinee J, Heijungs R, Hellweg S, Koehler A, Pennington D, Suh S (2009) Recent developments in life cycle assessment. J Environ Manage 91(1):1–21
21. Fischer TB (2002) Strategic environmental assessment in transport and land use planning. Earthscan Publications, London
22. Flénet F (2010) Lessons and limits of the study on the impact of the first generation biofuels coordinated by the French environment and energy management agency [Enseignements et limites de l'étude sur l'impact des biocarburants de première génération coordonnée par l'Ademe]. OCL—Oleagineux Corps Gras Lipides 17(3):127–132
23. Guinee J, Heijungs R, Huppes G, Zamagni A, Masoni P, Buonamici R, Ekvall T, Rydberg T (2011) Life cycle assessment: past, present, and future. Environ Sci Technol 45(1):90–96
24. Guinee J (ed) (2002) Handbook on life cycle assessment: operational guide to the ISO standards. Kluwer Academic Publishers, Dordrecht
25. IPCC (2012) Renewable energy sources and climate change mitigation. Special Report of the Intergovernmental Panel on Climate Change. Cambridge University Press. http://srren.ipcc-wg3.de/report/IPCC_SRREN_Full_Report.pdf. Accessed 17 sep 2013
26. Jeswani H, Azapagic A, Schepelmann P, Ritthoff M (2010) Options for broadening and deepening the LCA approaches. J Cleaner Prod 18(2):120–127
27. Jury C, Benetto E, Koster D, Schmitt B, Welfring J (2010) Life cycle assessment of biogas production by monofermentation of energy crops and injection into the natural gas grid. Biomass Bioenergy 34(1):54–66
28. Kaplan S, Garrick B (1981) On the quantitative definition of risk. Risk Anal 1:1–27
29. Kim H, Kim S, Dale B (2009) Biofuels, land use change and greenhouse gas emissions: some unexplored variables. Environ Sci Technol 43(3):961–967
30. Kuemmel B, Krüger Nielsen S, Sørensen B (1997) Life-cycle analysis of energy systems. Roskilde University Press, Roskilde
31. Melamu R, Blottnitz H (2011) 2nd Generation biofuels a sure bet? A life cycle assessment of how things could go wrong. J Cleaner Prod 19(2–3):138–144
32. Morris P and Therivel R (eds) (2009) Methods of environmental impact assessment (Second). SPON PRESS, London. Taylor & Francis Group. http://www.docstoc.com/docs/71241593/30592066-Methods-of-Environmental-Impact-Assessment. Accessed 17 sep 2013
33. Nieuwlaar E (2004) Life cycle assessment and energy systems. Encyclopedia of energy. Elvevier, pp 647–654
34. Rehl T, Lansche J, Muller J (2012) Life cycle assessment of energy generation from biogas—attributional versus consequential approach. Renew Sustain Energy Rev 16(6):3766–3775
35. Reinhard J, Zah R (2009) Global environmental consequences of increased biodiesel consumption in Switzerland: consequential life cycle assessment. J Cleaner Prod 17(Suppl 1):46–56
36. Reinhard J, Zah R (2011) Consequential life cycle assessment of the environmental impacts of an increased rapemethylester (RME) production in Switzerland. Biomass Bioenergy 35(6):2361–2373
37. Reisdorph D (2011) Rebound effects & monetizing environmental impacts. Paper presented at the Life Cycle Assessment (LCA) XI, October 4. Power Point. Chicago, IL
38. Sanden B, Karlstroem M (2007) Positive and negative feedback in consequential life-cycle assessment. J Cleaner Prod 15(15):1469–1481
39. Schmidt J (2010) Comparative life cycle assessment of rapeseed oil and palm oil. Int J Life Cycle Assess 15(2):183–197
40. Schmidt J, Weidema B (2008) Shift in the marginal supply of vegetable oil. Int J Life Cycle Assess 13(3):235–239

41. Silalertruksa T, Gheewala S, Sagisaka M (2009) Impacts of Thai bio-ethanol policy target on land use and greenhouse gas emissions. Appl Energy 86(Suppl 1):170–177
42. Soimakallio S, Kiviluoma J, Saikku L (2011) The complexity and challenges of determining GHG (greenhouse gas) emissions from grid electricity consumption and conservation in LCA (life cycle assessment)—a methodological review. Energy 36(12):6705–6713
43. Sokka L, Soimakallio S (2009) Assessing the life cycle greenhouse gas emissions of biorefineries. Paper presented at the VTT Symposium (Valtion Teknillinen Tutkimuskeskus). Technical Research Centre, Finland: VTT, pp 17–26. http://www.cabdirect.org/abstracts/20103325988.html;jsessionid=A0F426CA8D507015D9B687F35D37B9AF. Accessed 17 sep 2013 45.
44. Sovacool B, Bulan L (2013) They'll be dammed: the sustainability implications of the Sarawak Corridor of Renewable Energy (SCORE) in Malaysia. Sustain Sci 8:121–133
45. Thiesen J, Christensen T, Kristensen T, Andersen R, Brunoe B, Gregersen T, Thrane M, Weidema B (2008) Rebound effects of price differences. Int J Life Cycle Assess 13(2):104–114
46. Tillman A-M (2000) Significance of decision-making for LCA methodology. Environ Impact Assess Rev 20:113–123
47. Tukker A (2000) Life cycle assessment as a tool in environmental impact assessment. Environ Impact Assess Rev 20:435–456
48. Weidema B, Hischier R, Althaus H, Bauer C, Doka G, Dones R, Frischknecht R, Jungbluth N, Nemecek T, Primas A, Wernet G (2009) Code of practice. Swiss Centre for Life Cycle Inventories, Dübendorf. http://www.ecoinvent.org/fileadmin/documents/en/02_CodeOfPractice_v2.1.pdf. Accessed 17 sep 2013
49. Wolf M, Pant R, Chomkhamsri K, Sala S, Pennington D (2012) The International Reference Life Cycle Data System (ILCD) Handbook. Towards more sustainable production and consumption for a resource-efficient Europe. European Commission, Ispra. Joint Research Centre. Institute for Environment and Sustainability
50. Zamagni A, Guinee J, Heijungs R, Masoni P, Raggi A (2012) Lights and shadows in consequential LCA. Int J Life Cycle Assess 17:904–918

Chapter 4
Implementation of Hydrogen Gas as a Transport Fuel

Abstract There are high expectations for the use of hydrogen as a transport fuel in the future. However, as this chapter will show, sometimes these expectations are unrealistic, and are based on industrial actors' own agendas and strategies. The implementation of hydrogen energy in Norway has been heavily supported with governmental and industrial funding through the Hydrogen-road HyNor, but without expected advances in fuel cell technology and carbon capture and storage (CCS), the implementation has been rather limited. Residential opposition of hydrogen filling stations, because of health and safety concerns thwarted a pilot scheme in London, as will be shown. The life cycle GHG emissions from today's hydrogen fuels are high, in addition, the consequences from leakage of hydrogen gas from future production and distribution systems are potentially damaging to the stratospheric ozone layer.

4.1 Introduction

Hydrogen gas is an energy carrier which, at the moment, is largely produced by steam reforming of natural gas—a fossil energy. This production constitutes roughly 48 % of worldwide hydrogen [14]. However, hydrogen gas can also be produced from renewable energy, such as solar powered water electrolysis. Production through refining of biogas or gasification of biomass as another example. Hydrogen gas can thus, under these conditions, be produced with what is termed as *renewable hydrogen technology,* for example, in the technology domain definition of LCAs [3]. As such, hydrogen energy deserves mention in this book.

The implementation of hydrogen gas as a transport fuel can occur with the use of two different motor technologies:

1. Combustion of hydrogen gas in a combustion engine.
2. Consumption in a fuel cell to produce electricity that powers one or more electrical engines.

O. Andersen, *Unintended Consequences of Renewable Energy,*
Green Energy and Technology, DOI: 10.1007/978-1-4471-5532-4_4,
© Springer-Verlag London 2013

A key motivation for implementing hydrogen energy is its potential to reduce GHG emissions. However, this effect largely depends on the type of technology used in the production of hydrogen [15]. The reported reductions in GHG emissions from using a hydrogen fuel cell vehicle instead of a comparable gasoline vehicle varies from 100 % reduction to a 60 % increase depending on the hydrogen gas production technology [19]. It is a paradox that fuel cell vehicles using hydrogen produced with some of the "greenest" technology options for hydrogen production—electrolysis of water—consume more energy than gasoline vehicles. This is highly problematic from an energy-efficiency point of view, where the aim is to reduce energy use, not increase it.

Studies on the non-technological barriers for hydrogen use in transport have revealed a number of unintended consequences for the implementation of this energy carrier. The RENERGI ("Clean energy of the future") program at the Research Council of Norway (RCN) funded the research project: "Providing hydrogen for transport in Norway: A social learning approach." The project was coordinated by Western Norway Research Institute (WNRI) in collaboration with the Department of Interdisciplinary Cultural Studies at the Norwegian Technical and Natural Science University (NTNU) in the period 2005–2010. The research included an evaluation of the hydrogen implementation project "The hydrogen road Oslo-Stavanger" (HyNor).

4.2 The Hydrogen Road: HyNor

The HyNor hydrogen road from Oslo to Stavanger was presented as a connected infrastructure network facilitating the use of hydrogen as a transport fuel, with demonstration of various technologies for producing hydrogen at the various nodes. A map of HyNor with its nodes indicated (as it was in 2005) is shown in Fig. 4.1.

Fig. 4.1 The hydrogen road HyNor with its nodes indicated

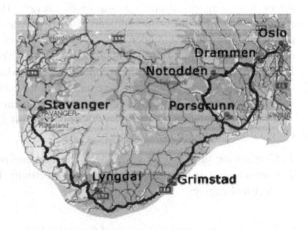

In his doctoral thesis, Kaarstein [10] points to a series of unintended consequences of hydrogen-implementation in Norway, which were revealed through the
evaluation of the HyNor project. He characterizes HyNor as a project aiming to
simulate a range of possible future technology projects – i.e., a meta-technology
development project.

Given a failed attempt to include Oslo as a demonstration city in the EU-funded
project Clean Urban Transport for Europe (CUTE), the large Norwegian industrial
company Hydro launched the project with the name "The Hydrogen Road of
Norway." The diversity of the various hydrogen technologies in the demonstration
cities in CUTE were used as inspiration for modeling HyNor. The intention was
that the HyNor cities (nodes) would test and demonstrate different technologies for
production and use of hydrogen as a transport fuel. The heterogeneous character of
the nodes in HyNor emerged as a key feature of the project.

The implementation of hydrogen through HyNor would soon face a series of
unexpected issues, conflicts and obstacles. It proved to be rather difficult to arrive
at viable and credible concepts for hydrogen production—including carbon capture and sequestration/storage (CCS). The construction of an adequate and comprehensive infrastructure was also problematic as was the simple availability of
suitable hydrogen cars, complete with fuel cells.

Given all this it might be argued that this clean energy implementation project
was nothing more than "hype". But, instead of being vacuous, Kaarstein states that
there were several indications that it simply was too early for the HyNor actors to
initiate a project of this type. From its start in 2003, the necessary technology was
not emerging fast enough. In particular, it appeared that the development of
energy-efficient hydrogen cars was much slower than the key players had hoped
for. This leads Kaarstein to state that HyNor did demonstrate in a fairly efficient
way that there is a long way to go before we can enter a "hydrogen society."

An unintended consequence of hydrogen implementation, as exemplified by
HyNor, is that governmental funding, matched with industrial funds, sometimes
are spent on immature energy technologies. There are many reasons for this, but
one became quite clear in the HyNor project; the technologies were promoted by
stakeholders who had their own interests in the realization of the technologies,
even though the realization potentials were quite unrealistic. Hydro aimed to
deliver the electrolyzers for hydrogen production to several of the nodes in the
project, as they had done to several of the CUTE-cities. Hydro was also a major
producer of hydrogen and wanted to use HyNor to increase sales of hydrogen gas.
Statoil hoped to demonstrate hydrogen production from its natural gas, combined
with CCS. All these plans proved overly difficult to implement for various reasons.
For example the car manufacturer Think entered into bankruptcy before their
hydrogen cars were produced. This is dealt with in detail in Chap. 6.

4.3 Clean Urban Transport for Europe

The research program "Systems for the implementation of hydrogen energy in the transport sector," was funded by RCN and conducted by WNRI. A key task was to utilize and advance knowledge from the implementation of hydrogen in countries other than Norway. Specific knowledge was summarized and systematized from the project Ecological City Transport System (ECTOS) in Reykjavik, and the CUTE-project, which encompassed Amsterdam, Barcelona, Hamburg, Stockholm, Porto, Stuttgart, London, Madrid, and Luxembourg. Knowledge and experience was also gathered from the project HySociety,[1] on the importance of assessing unintended consequences and barriers [1, 2].

A crucial unintended consequence that delayed the demonstration of hydrogen use in hydrogen buses in London was public opposition among local residents to the localization of the hydrogen filling station. Protests from neighbors in the vicinity resulted in significant delays, so that the process toward approval, which began in 2001, did not come to an end until November 2004. It became obvious that such opposition had not been expected by the promoters of CUTE and thus, not taken into consideration in the planning process. The complaints from the public included:

- The site of the planned filling station was on an existing green area used by local residents for recreational purposes
- The site had large negative visual impact with tall fences and lights
- Fear of accidents. The hydrogen filling station had in the local press been associated with the Hindenburg[2] accident and hydrogen bombs.

The delay in the approval of the location for the hydrogen filling station led to the use of a temporary filling station for the start of the bus demonstration in January 2004. A key lesson from the hydrogen implementation in London was that unintended consequences in the form of public opposition must be taken into consideration when planning hydrogen infrastructure.

---

[1] HySociety was a project ("Accompanying measure") funded by the EC in the 5th framework program for research and development. The project identified non-technological barriers, and evaluated technologies and policy measures for establishing a European hydrogen economy. The basis was knowledge reviews on social barriers, technologies, and other relevant aspects of social, economic, and environmental nature. In addition to analyzing conditions and barriers, measures to overcome the identified barriers were recommended [4].

[2] The hydrogen-filled airship (zeppelin) "Hindenburg" exploded in the air on 6 May 1937 killing 35 people.

4.4 Technical Spillover Effects of Energy Policies

Hertwich [7] points to hydrogen energy in connection with the rebound debate and unintended consequences of hydrogen policies in connection witht thier ability to achieve their goals. The key issue he raises is that technical changes necessary for utilization of hydrogen energy can enable other emission-reducing technologies. In an analysis of hydrogen fuel cell vehicles, it was found that such vehicles are environmentally preferable to hybrid combustion engine-electric cars in Europe, but not the United States, because electricity is cleaner in Europe [8]. This is based on the presupposition that it is likely that a hydrogen distribution system would use more electricity than fossil fuel distribution systems do [7]. This is termed a *technical spillover*, where cleaner electricity favors cleaner transportation technology, in this case based on hydrogen energy.

4.5 Stratospheric Ozone-depletion

The impacts of a future hydrogen economy include potential consequences for the ozone-layer in the stratosphere [18]. The unintended emission of molecular hydrogen (H_2) from future widespread use of hydrogen as fuel has the potential to significantly increase the abundance of water vapor in the stratosphere. The water increase, plausibly as much as 1 part per million (ppm), would cause stratospheric cooling, enhancement of the heterogeneous chemistry that destroys ozone, plus an increase in noctilucent clouds.

This reasoning has basis in assumptions on leakage of molecular hydrogen from infrastructure and operations connected to the production, distribution, and use of the hydrogen fuel. Both stationary and mobile applications of hydrogen gas would contribute. In principle, a perfectly efficient system of hydrogen infrastructure and use would involve no H_2 leakage. In that case, the move away from fossil fuel combustion to hydrogen energy would actually result in a reduction in anthropogenic H_2 emissions, because fossil fuel combustion results in H_2 emissions.

However, the amount of molecular hydrogen in the atmosphere is increasing. An average annual increase of 0.6 % between the years 1985 and 1989 has for example been reported [12]. In addition, based on experiences with technologies connected to production and transportation of natural gas and other volatile products, it seems likely that systems for H_2 production, storage, and distribution will involve losses to the atmosphere. The magnitude of these losses depends on the efforts that are made to avoid them, but estimates have been projected to about 10 % [21]. Due to the fact that losses during commercial distribution of H_2 are substantially greater than this [16], a range of 10–20 % can be expected [18]. Then, if in the future, all current technologies based on fossil diesel and gasoline were replaced with hydrogen technologies, the anthropogenic emissions of H_2 would be between 60 and 120 Tg/year. This is in the range of four to eight times the estimates for the current anthropogenic H_2 emissions, which is 15 ± 10 Tg/year. In that case, the contributions from human activities would dominate, and result in a doubling or tripling of the yearly production of H_2 from all sources combined.

Due to its extremely high volatility, molecular hydrogen easily moves up from the lower part of the Earth's atmosphere (the troposphere), and through the tropopause at about 10 km altitude. Then it continues upwards into the stratosphere where it is oxidized to water [13]. The moistening of the stratosphere would result in a cooling of its lower parts [6]. This is where the protective ozone-layer is located, and the increase in moisture would disturb the chemistry involved in its recovery after the massive destruction primarily caused by human releases of chlorofluorocarbons (freon gases, CFCs). The stratospheric ozone layer is maintained through a balanced set of chemical reactions. Nitric acid in polar stratospheric clouds reacts with CFCs to form chlorine, which catalyzes the photochemical destruction of ozone. These reactions are effected by, among others, moisture and temperature [17]. The increase in stratospheric water content caused by a quadrupling of the anthropogenic emissions of H_2 would have the consequence of lower stratospheric temperatures [18]. Colder temperatures would create more polar stratospheric clouds, delay the breakup of the polar vortex, and thereby make the ozone hole wider and more persistent in the spring.

Thus, the future anthropogenic emissions of molecular hydrogen in a hydrogen economy could substantially delay the recovery of the ozone layer that is expected to result from the regulations for phasing out CFCs [11, 20, 22]. Additionally, the modeling by Tromp et al. [18], of the consequences from anthropogenic hydrogen losses, gave a potential 10 % increase in the stratospheric concentration of the hydroxyl (OH) radicals, which reacts with ozone to produce oxygen and peroxy (HOO) radicals. Thus an increased breakdown of stratospheric ozone can be expected. In addition, the increased amounts of H_2O derived from H_2 could lead to an increase in noctilucent clouds, with potential impact on the Earth's albedo effect and the chemistry of the mesosphere (the layer of the atmosphere right above the stratosphere). Also, since H_2 is a microbial nutrient, the increased presence of H_2 over soil systems might have unforeseen effects on microbial communities [18].

Hydrogen leakages could also have implications for climate, as recognized by the International Panel on Climate Change. In their third assessment report [9], it is pointed out that hydrogen can negatively interfere with the atmospheric chemistry responsible for abating methane and other major greenhouse gases. It clearly states that in a possible fuel cell economy, future (hydrogen) emissions may need to be considered as a potential climate perturbation [9].

4.6 Other Environmental Risks in a Hydrogen Economy

Flynn et al. [5] call for comprehensive assessments of health and environmental risks, which take into account the whole life cycle of the technological system of which hydrogen will be a part of. A wide array of established and new technologies will contribute to the production, storage, distribution, and use of hydrogen. Materials such as metal hydrides, carbon nanotubes, and various catalysts will be variably deployed across the hydrogen energy chain, in amounts that will depend upon the scale of hydrogen penetration in the economy and the relative adoption rates of different hydrogen technologies. Increased production, diffusion, and disposal of such materials, some of which may be totally newly engineered, may have risk implications for public health and the environment.

References

1. Andersen O (2006) Ikke-teknologiske barrierer for hydrogen som energibærer i transport. Hva kan vi lære av CUTE, ECTOS og HySociety? (Non-technological barriers for hydrogen as energy carrier in transport. What can we learn from CUTE, ECTOS and HySociety?). Western Norway Research Institute: Sogndal. http://www.vestforsk.no/ filearchive/notat11-06-hydrogen.pdf. Accessed 17 sep 2013
2. Andersen O (2007) Hydrogen as transport fuel in Iceland. The political, technological and commercial story of ECTOS. Int J Altern Propul 1(4)(4):339–351
3. Chen I-C, Fukushima Y, Kikuchi Y, Hirao M (2012) A graphical representation for consequential life cycle assessment of future technologies - Part 1: methodological framework. Int J Life Cycle Assess 17:119–125
4. European Commission (2005) HySOCIETY 'An innovative (accompanying) measure aiming to support the introduction of a safe and dependable hydrogen-based society in Europe'. Power Point
5. Flynn R, Bellamy P, Ricci M (2006) Risk perception of an emergent technology: The case of hydrogen energy. Forum: Qualitative Social Research/Forum: Qualitative Sozialforschung 7(1): Art. 19. http://nbn-resolving.de/urn:nbn:de:0114-fqs0601194. Accessed 17 sep 2013
6. de Forster P and Shine K (2002) Assessing the climate impact of trends in stratospheric water vapor. Geophys Res Lett 29(6):10–1—10–4
7. Hertwich EG (2005) Consumption and the Rebound Effect: An Industrial Ecology Perspective. J Ind Ecol 9(1–2):85–98
8. Hertwich EG and Strømman A (2004) The environmental benefit of direct hydrogen fuel cell vehicles. An analysis of the assessment literature. Norwegian: Trondheim, Norway. University of Science and Technology, Industrial Ecology Programme
9. IPCC (2001) Third Assessment Report (TAR) of the Intergovernmental Panel on Climate Change (IPCC) prepared by Working Group III: Climate Change 2001: Mitigation. Intergovernmental panel on climate change. http://www.ipcc.ch/ ipccreports/tar/wg3/index.php?idp = 144. Accessed 17 sep 2013
10. Kaarstein A (2008) HyNor - den norske hydrogenveien? En studie av en stor tekno-politisk hybrid (HyNor - the Norwegian hydrogen road? A study of a large techno-political hybrid). PhD Thesis, Trondheim, Norway, Norges teknisk-naturvitenskapelige universitet. http:// ntnu.diva-portal.org/smash/get/diva2:139432/FULLTEXT01. Accessed 17 sep 2013
11. Kalnay E, Kanamitsu M, Kistler R, Collins W, Deaven D, Gandin L, Iredell M, Saha S, White G, Woollen J, Zhu Y, Leetmaa A, Reynolds R, Chelliah M, Ebisuzaki W, Higgins W, Janowiak J, Mo K, Ropelewski C, Wang J, Jenne R, Joseph D (1996) The NCEP/NCAR 40-Year Reanalysis Project. Bull Am Meteorol Soc 77(3):437–471
12. Khalil M, Rasmussen R (1990) Global increase of atmospheric molecular hydrogen. Nature 347:743–745

13. Letexier H, Solomon S, Garcia R (1988) The Role of Molecular-Hydrogen and Methane Oxidation in the Water-Vapor Budget of the Stratosphere. Quarterly Journal of the Royal Meterological Society 114(480):281–295
14. Pehnt M (2003) Life-cycle analysis of fuel cell system components. *Handbook of fuel cells— fundamentals, technology and applications.* Wiley & Sons, Ltd: Chichester, pp 1293–1317
15. Sanden B, Karlstroem M (2007) Positive and negative feedback in consequential life-cycle assessment. Journal of Cleaner Production 15(15):1469–1481
16. Sherif S, Zeytinoglu N, Veziroglu T (1997) Liquid hydrogen: Potential, problems, and a proposed research program. Int J Hydrogen Energy 22(7):683–688
17. Solomon S (1999) Stratospheric ozone depletion: A review of concepts and history. Rev Geophys 37(3):275–316
18. Tromp T, Shia R-L, Allen M, Eiler J, Yung Y (2003) Potential environmental impact of a hydrogen economy on the stratosphere. Science 300(5626):1740–1742
19. Wang M (2002) Fuel choices for fuel-cell vehicles: well-to-wheels energy and emission impact. J Power Sources 112(1):307–321
20. WMO (2011) Scientific assessment of ozone depletion: 2010. Global ozone research and monitoring project. World Meteorological Organization: Genova. US National Oceanic and Atmospheric Administration, US National Aeronautics and Space Administration, United Nations Environment Programme, European Commission
21. Zittel W and Altman M (1996) Molecular Hydrogen and Water Vapour Emissions in a Global Hydrogen Energy Economy. In: Veziroglu T, Winter C-J, Baselt J and Kreysa G (eds) Proceedings of the 11th World Hydrogen Energy Conference. Stuttgart, Germany: Schön & Wetzel, Frankfurt am Main, Germany, 71–82
22. Zurek R, Manney G, Miller A, Gelman M, Nagatani R (1996) Interannual variability of the north polar vortex in the lower stratosphere during the UARS mission. Geophys Res Lett 23(3):289–292

Chapter 5
Biodiesel and its Blending into Fossil Diesel

Abstract This chapter examines the various unintended consequences of biodiesel production including loss of biodiversity and shortfalls in GHG reductions. An account of biodiesel impacts in comparison to fossil diesel is followed by a critical review of the common practice of using additives to improve biodiesel performance in the winter. Many of these additives have negative, or unknown effects on human health and the environment. In this respect, biodiesel is an environmentally "friendly" fuel that creates environmental problems. An additional aspect of biodiesel use is the practice of blending the fuel with fossil diesel. Common in Europe, USA, and Canada, blending is done in order to comply with policy targets for increasing the share of transport fuels based on renewable energy sources. Results obtained from advanced modeling, designed to predict future consequences of the blending practice are then presented in this chapter. These include molecular dynamics simulations indicating that new toxic nanoparticles are being formed in the exhaust pipes of vehicles run on bio-blended diesel. This represents a likely mechanism for the increased exhaust mutagenicity of bio-blended diesel observed in other studies.

5.1 The Unintended Consequences of Biodiesel Production

First-generation biofuels[1] have been shown to have negative impacts on biodiversity, ecosystems, climate, food security, and the inclusion of the poor [9]. In the case of biodiesel, a key unintended consequence has been the massive eradication of indigenous vegetation or existing food crops to make way for mono-culture palm plantations in order to generate sufficient palm oil [36]. This has already

---

[1] First-generation biofuels are made from "sugar, starch and oil bearing crops or animal fats that in most cases also can be used as food and feed", according to the International Energy Agency [13].

O. Andersen, *Unintended Consequences of Renewable Energy,*
Green Energy and Technology, DOI: 10.1007/978-1-4471-5532-4_5,
© Springer-Verlag London 2013

reduced biodiversity significantly in the areas affected, including swamps, peat land, and tropical rain-forests in Indonesia, Malaysia, and Thailand [27, 41].

The draining of land areas for biofuel production has also effected the emission of GHGs. When the swamps in Indonesia were drained to establish large-scale palm oil production, there was an acceleration of soil decomposition, unexpectedly releasing large quantities of GHGs into the atmosphere [43, 48]. Several European nations, including France and Germany, withdrew support for palm oil when it was discovered that the increased GHG emissions were substantially greater than the potential GHG reductions from the fossil fuel to biodiesel transition.

Critical assessment of biofuels and their implementation is necessary to avoid the generation of unexpected negative consequences. This includes new knowledge on how they impact on the environment, and the world's food supply. This is an area of knowledge generation that can be facilitated through barrier studies, where the various factors that hinder implementation are studied. International projects made important contributions to this type barrier knowledge.

Barrier studies have frequently been funded by the Intelligent Energy Europe (IEE), EU Competitiveness and Innovation Programme (CIP) and also occasionally by the European Union's Framework Programme (EU-FP) for research and technological development (RTD). Crucial knowledge was acquired through the IEE sub-programme for alternative energy (Altener), including the project "Biodiesel in heavy-duty vehicles, strategic planning and vehicle fleet experiments". In this project, the barriers were revealed, and provided a basis for developing a strategic plan for the use of biodiesel in heavy vehicles [2]. The project, conducted between 1996 and 1998, revealed a series of unintended consequences from the large-scale implementation of biodiesel. Central to the project was the issue of large greenhouse gas emissions from the production phase of the life cycle of biodiesel. It was shown that in a "worst case" scenario, based on intensive and large-scale agricultural systems, a transition from fossil diesel to the use of biodiesel could actually increase GHG emissions by 15 %.

The main reason for the high GHG emissions for biodiesel production is that intensive cultivation of rape (*Brassica nupus*), the most common oil seed crop for biodiesel in Europe, is only possible with the use of large amounts of nitrogen-rich artificial fertilizers. This results in the release of N_2O (nitrous oxide), which is a strong GHG with a global warming potential (GWP) of 310 times that of CO_2 in a 100-year time horizon [8]. Taking this emission into consideration was a key factor in lowering the predicted GHG reduction from replacing fossil diesel use with biodiesel produced in Norway [1, 2]. The life cycle GHG reduction was previously estimated at 60 %, but with the inclusion of N_2O-emissions this was lowered to 17 %. The "highest" estimate (+50 % uncertainty in the calculation of the N_2O-emissions) actually gave a 15 % increase in greenhouse gas emissions from the replacement of fossil diesel by rape seed methyl ester (RME).

In addition it has been known for several decades that N_2O also has a damaging effect on stratospheric ozone, and as such is classified as an ozone-depleting gas [4].

This unintended consequence from the implementation of a renewable energy source was not discussed further until a decade later, when it became a part of the biofuel debate and central to the quality requirements in effect today, e.g through the sustainability criteria of the European Commission [7]. These criteria specify that sustainable production of biofuels requires that raw material is not grown in areas that are important for biodiversity or in soil with high fixed carbon content, such as wetlands. The production of biofuels must prove a significant reduction in GHG emissions by at least 35 %, in comparison with fossil fuels. In 2017 this demand for emission reduction will increase to 50 %.

5.1.1 Additive Usage

A group of unintended consequences from biodiesel implementation stems from the increased use of additives to achieve fossil fuel levels of performance. Additive usage in biodiesel made from rape seed (rape seed methyl ester, RME) has been reviewed [2]. Indications of environmental impacts connected with the various additive types were included. Although somewhat dated and as such not representing a complete overview of additives in use, the review points to the major problem of obtaining proprietary information on the additive compositions, in order to assess their unintended impacts in the form of health and environmental implications. A more recent review was made by Ribeiro et al. [39], in which the need for additive use in combination with biofuels, are substantiated to combat the poor fuel qualities of biofuels. These disadvantages include higher NOx emissions, shorter intervals between motor parts replacement, such as fuel filters and deposits in fuel tanks and fuel lines, etc.

Most additives marketed for use in biodiesel are mixtures of different compounds blended together into additive packages, to provide a number of functions simultaneously. Substantial research has been conducted in improving the properties of fuels by finding *combinations* of different types of additives [47]. The environmental effects of additive usage are not limited to the effects of individual compounds. The possibility for *synergistic effects* of each individual compound must also be taken into account when assessing the unintended consequences of additives on health and environment.

The main reason for using additives in biodiesel is to improve the winter properties of the fuel. This is necessary because at temperatures below −5 °C the plugging of fuel lines and filter is common. Two different terms are used to describe the cold flow properties. CCFPP is an abbreviation both for *critical cold filling pouring point* and for *critical cold filter plugging point*. With a CCFPP of −20 °C the fuel is suitable for use at temperatures down to −20 °C. Additives which increase the cold flow properties, by lowering the CCFPP, are termed *pour point depressors* (PPD).

Substantial research on cold flow additives has been conducted at the Northern Agricultural Energy Centre and National Centre for Agricultural Utilization Research (NCAUR) in Peoria, Illinois, USA. They have studied biodiesel from soy bean, and the physical chemistry involved in additive treatment is analogous for other types of biodiesel as well.

Nearly, all of the additives tested by NCAUR were developed and marketed for treatment of conventional diesel fuel (petroleum middle distillates). Most of these additives have active compounds such as ethylene vinyl acetate copolymers, alkenyl succinic amides, high molecular weight long-chain polyacrylates, fumarate-vinyl acetate copolymers, and copolymers of linear alpha-olefins with acrylic, vinylic, and maleic compounds. These additives also, typically contain a petroleum-based solvent or "vehicle" such as aromatic naphtha. Naphtha can lead to cancer on contact with skin and it can cause airway irritations and respiratory problems. It can even lead to coma if present at high concentration. Chronic exposure can cause headache, reduced appetite, dizziness, sleeplessness, indigestion, and nausea [28].

The focus on additives for improving biodiesel properties has also resulted in attempts to develop new additives based on biological resources. Experiments have been conducted, blending medium-long chain alcohols and methanol [5, 20–22]. The results of these experiments confirm that none of these additives will lower the CCFPP better than by mixing in fossil diesel. Similar conclusions have been reached at the research group at Department of Biological and Agricultural Engineering at University of Idaho. The mixing in of paraffin in cold weather, in addition to additive usage, is a common practice. An unintended consequence of this blending is that it can lower the cetane number of the fuel, thereby requiring other additives to improve the cetane number again.

Lubrizol International Laboratories is marketing PPD-additives based on three main structures shown in Figs. 5.1, 5.2 and 5.3.

When $R = CH_3$ in Fig. 5.2 this is polymethylmethacrylate, also known as the material "liquid Plexiglas", a suspected carcinogen [28].

Ethylenevinylacetate (Fig. 5.3) copolymer, will upon combustion degrade into a large number of different straight-chain hydrocarbons, and add to air pollution in the same way as other volatile organic compounds do [31]. The combustion gas from this additive is irritating to skin, eyes, and the mucous membranes of the nose and throat [28].

At the Belgian research institute VITO (Vlaamse Instelling voor Technologisch Onderzoek) the C_8–C_{10} fraction (the lightest fraction) of coco methyl ester (CME), produced by Fina Oleochemicals in Belgium, has been used to overcome the cold temperature problems associated with biodiesel use. By using a 20 % blend of CME in biodiesel, a lowering of CCFPP down to -15 °C was achieved in tests

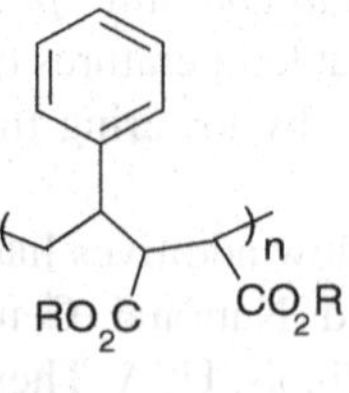

Fig. 5.1 Molecular depiction of Melan-styrene ester

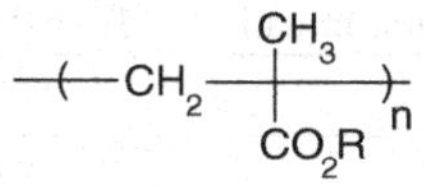

Fig. 5.2 Molecular depiction of Polymethacrylate

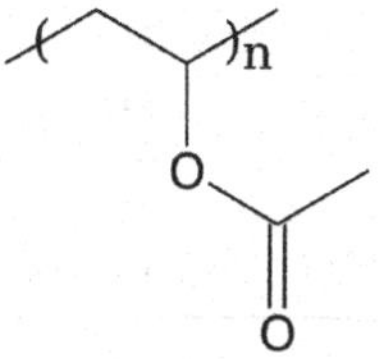

Fig. 5.3 Molecular depiction of Ethylenevinylacetate

[18]. The problem with smell from the exhaust was however increased when this mixture was used. VITO did not continue with these investigations, in part due to the long distances and transport costs of the coco oil from the tropical regions to the place of usage (in cold weather regions).[2]

5.1.2 Alternatives to Additive Usage

In addition to using additives, other methods have been developed to improve the winter properties of biodiesel. Terms like "winterizing" and "de-clouding" refer to the removal of hight-melting fatty acid methyl esters (FAME), from the bio-diesel. This is done by slowly cooling down the biodiesel to a critical temperature where the high molecular FAME are precipitated out of solution and settled at the bottom. The sediment is removed and used in the summer as a fuel with a higher CCFPP. A special winter biodiesel with a CCFPP of –38 °C can be obtained using this method of physical separation. This is, however, only possible in combination with an additive [38]. There is a serious unintended consequence of this alternative to additive usage, namely that the process of physical separation requires substantial amounts of energy input. The fatty acid composition in this winter-RME and native biodiesel is shown in Table 5.1 and demonstrates that a significant proportion of the $C_{16:0}$ methyl ester is reduced.

Another aspect of biodiesel that can be improved is the deposit-forming tendency due to polymerization of the fuel. This property is partially determined by the degree of unsaturation in the fatty acids. The polymerization tendency of the fuel goes down if the number of unsaturations is reduced. Low volatility of the polymerised fuel causes it to be washed down along the cylinder walls and end up in the engine oil,

---

[2] Information obtained through personal communication with A. Demoulin, Fina Oleochemicals, Fina Research S. A., Zone Industrielle C, 7181 Seneffe (Feluy), Belgium. Tel.: +32/64/51.42.29.

Table 5.1 Main fatty acid composition in native RME and special winter–RME

Fatty acid	% mass		
	Native RME	Winter-RME	HORO
$C_{16:0}$	4	1	4
$C_{18:0}$	1	–	2
$C_{18:1}$	60	61	72
$C_{18:2}$	22	25	15
$C_{18:3}$	10	11	2
$C_{20:0}$	–	–	1
$C_{20:1}$	1	2	–
$C_{20:2}$	–	1	1

thereby diluting it. To reduce this problem, new hybrids of the rape plant are being developed by breeding and genetic engineering. For example, a "high oleic rape seed oil" (HORO), with iodine number of 100, as opposed to 118 for normal rape, is currently commercially available [26]. Genetically engineered rape plants as raw materials for production of biodiesel with improved properties are being developed by many industrial companies, among them Ciba-Geigy.

The development of new plant properties by the use of genetic engineering can however represent a conflict with the *precautionary principle*, as it is impossible to guarantee that no unpredicted irreversible environmental effects from this technology will appear in the future. The precautionary principle states that if there is such uncertainty for irreversible effects on the environment, the lack of full scientific proof for the effect is not a good enough argument for not implementing actions to reduce the effects. The conflict with this principle is further dealt with in the next paragraph.

5.2 Unintended Consequences of the Blending Strategy

Another group of unintended consequences of biofuels stems from the way the fuel is implemented into the transport sector. The phasing-in of biofuels is to a large extent conducted through blending biofuels into the existing fossil fuel sold at filling stations. Alcohol is blended into gasoline (e.g. E85, i.e. fuel consisting of 15 % ethanol and 85 % gasoline), while biodiesel is blended into regular diesel (e.g. B7, consisting of 7 % biodiesel and 93 % fossil diesel). This strategy for increasing the use of biofuels is common in large parts of the EU, in the United States and in Canada. The European Standard (EN590) for automotive diesel fuel states that the concentration of biodiesel in diesel fuel should be 7 % v/v. This concentration is expected to rise up to 10 %, based on European policy targets, including the "20-20-20" strategy "EUROPE 2020 – A Strategy for Smart, Sustainable and Inclusive Growth" by the European Commission [6, 23].

Unintended consequences of the bio-blending strategy has been studied by a variety of research groups [11, 17, 24, 33, 46]. The topic was also studied in the European Economic Area (EEA) / Norway Grant project "Influence of bio-components content in fuel on emissions from diesel engines and engine oil deterioration–BIODEG" [35].

Through the BIODEG project it was shown that there are some very specific, problematic, toxicological issues that can result in unexpected negative consequences from the biodiesel blending strategy [3]. The formation of new types of exhaust emissions were substantiated by the use of molecular dynamic simulation (MDS) studies in the KTH supercomputer facility in Stockholm. The studies were conducted on phenanthrene (Phe), which is a polynuclear aromatic hydrocarbon (PAH), and Oleic Methyl Ester (9-cis-octadecenoic methyl ester – OME). Phe was chosen as a reference molecule for three-ringed PAHs, because it is the dominant component among combustion-generated PAHs [32, 37], and is of ecotoxicological concern [29]. The molecular coordinates of Phe were collected from the Hic-Up database [19] and protonated. The charges and atom types were assigned according to the OPLS/AA force field [15, 16] in the GROMACS molecular dynamics package [12, 44 OME was used as a model for FAMEs in biodiesel fuel, as it is present in high concentrations in most common types of biodiesel. The structure of the OME molecule was simulated using the PRODRG server [42], while the atom types and charges were assigned accordingly with the OPLS/AA force field in the GROMACS package. The TIP4P water model [14] was used throughout the simulations.

Through these studies it was shown that new types of nanoparticles can be formed through an aggregation process. This is shown in Fig. 5.4, where aggregation of OME and Phe is observed after short simulation.

The gradual formation of nanoparticles is also seen in Fig. 5.5, where three consecutive snapshots were taken.

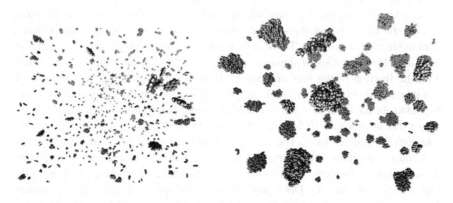

Fig. 5.4 Aggregation of PAH (Phe) and FAME (OME) into nanoparticles. *Left* initial configuration, *right* after short simulation

Fig. 5.5 Snapshots of a mixture of FAME (OME) and PAHs (Phe) at increasing lapsed time (0, 10, 45 s) after the start of the simulation. The FAME-molecules are shown in *blue* color, the PAH in *pink*

The sequentially formed nanoparticles simulated and depicted in Figs. 5.4 and 5.5 are composed of FAME and PAHs, have polar moieties, constituting the methyl ester head of the FAME molecules (red color in Fig. 5.6b).

The property of FAME and PAH molecules that are crucial for their abilities to form PAH-FAME agglomerates is the distribution of electron density/charge throughout the individual molecule. The charge distribution of FAME, shown in Fig. 5.6, is not symmetrical, but with a dipole, with the ester moiety constituting an electron surplus region, i.e., a polar "head" and a nonpolar (aliphatic) hydrocarbon "tail". The PAH molecule, on the other hand, has a symmetrical electron density distribution and is thus a nonpolar molecule. Due to these different properties, the FAME-molecules orientate their polar head proximate to the water surface and their tail inward away from water. The PAH has a tendency to be present mainly near the nonpolar hydrocarbon tail of the FAME molecules. The moisture in the air enables the PAH to be carried, due to the polar FAME-head's affinity for water. Membrane penetration could be facilitated, resembling "flip-flop" mechanisms, due to the polar moieties with their affinity for membrane proteins and the phosphate head of the phospholipids, and the nonpolar moieties' affinity for the aliphatic segments of the phospholipids in the membrane.

With the polar moieties of the FAME molecules, and the PAH presence in these predicted nanoparticles, it is plausible that they can alter the exhaust composition, making the exhaust from bio-blended diesel more carcinogenic than the exhaust from unblended diesel [30]. This can be understood when taking into consideration the particular properties of these nanoparticles. They are plausibly being formed from the fraction of the fuel that passes through the combustion system more or less intact. This fraction is quite high, particularly for the biodiesel component of the blend, in cold weather and and during "cold starts" in general. Studies have shown that as much as 10 % of all emissions in the United States derive from the cold-start of engines [41]. The non-combusted (or partially combusted) FAME molecules, with their polar "heads" can function as a "vehicle" for transporting the nonpolar carcinogenic PAHS across the cell membranes and into the lung cells, where they can bind to DNA and cause mutations, and thus lead to cancer development. Simulations of these PAH-FAME nanoparticles are shown in Figs. 5.7 and 5.8.

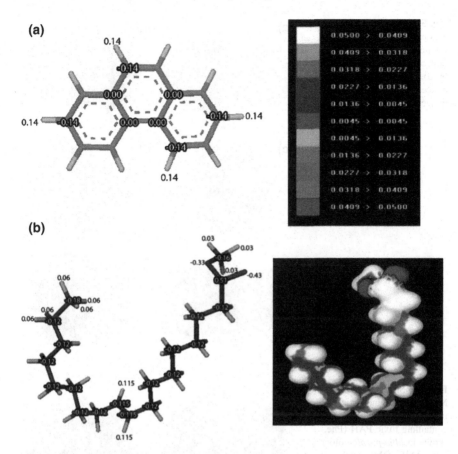

Fig. 5.6 Atomic charges and electron density of Phe (**a**) OME (**b**). To the *left* the molecules are shown with the assigned atomic charges. The charges derive from the OPLS/AA force field included in the GROMACS molecular dynamics simulation package. All hydrogen atoms (shown in *white*) bound to sp3 carbon atoms between the methyl end and the double-bonded (diene) carbons, and between diene carbons and the ester group, are given the charge 0.06, but this is not shown in the figure. Oxygen atoms are shown in *red*. In the electron density plots (to the *right*) the atomic charges are visualized in colors according to the legend (far *right*)

The internal structure of the simulated PAH-FAME nanoparticles was characterized through quantifying the relative density of the components in the nanoparticles, in a radial fashion. The results are shown in Figure 5.9 a, b and c. At low Phe concentration this molecule prefers the exterior part of the particles, while at higher Phe concentrations the molecules are spread out more. The OME head groups tend to cluster somewhat on the outside, even in vacuo. In water the picture is slightly different. The OME head groups cluster much more strongly at the surface, in a manner resembling the molecular orientations in micelles [45]. The Phe molecules prefer to concentrate slightly below the surface, while the OME

Fig. 5.7 Simulated nanoparticles in vacuum, each consisting of 512 OME molecules, plus 10 (**a, b**), 50 (**c, d**), and 100 (**e, f**) Phe molecules. The Phe molecules are shown in *blue*

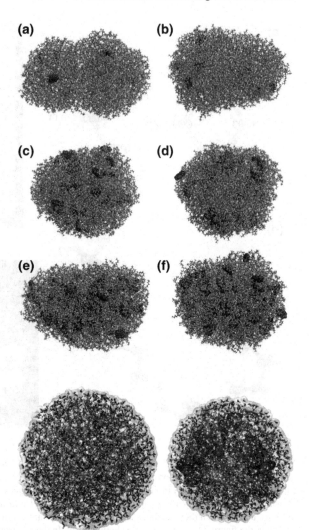

Fig. 5.8 Snapshots from a MDS of two nanoparticles immersed in water. *Right* containing toxic PAH (Phe, shown in *blue-purple* color) and FAME (OME), *left* containing OME only. The carbon atoms are shown in *green*, oxygen in *red*, hydrogen in *white*

tails make up the core of particle. Notably, the inner parts of the particle are wet: there is water at low density, dragged inside by the OME head groups.

The results of these MDS studies indicate that nanoparticles can be formed from non-combusted FAME. These nanoparticles are capable of incorporating PAHs. Moisture in the air can carry the nanoparticles, which constitute a vehicle for the PAHs. They can then be carried into the lungs where they can penetrate the cell walls, enter the interior of the cells, and interact with DNA. This can potentially initiate carcinogenic activity. A model of such a nanoparticle is shown in contact with a phospholipid bilayer membrane (main structural constituent of cell walls) in Fig. 5.10.

Fig. 5.9 Radial distribution functions of the agglomerates in vacuo (**a–c**) and in water (**d**). **a** 10 Phe molecules and 512 OME molecules, **b** 50 Phe molecules and 512 OME molecules, **c** 100 Phe molecules and 512 OME, molecules and **d** 22 Phe molecules and 76 OME molecules. The radial distribution function describes the stability of the molecules into chemically stable nanoparticles, including the orientation of the respective molecular groups. OME head represents the ester group of the OME molecules, while OME tail represents the hydrophobic moiety

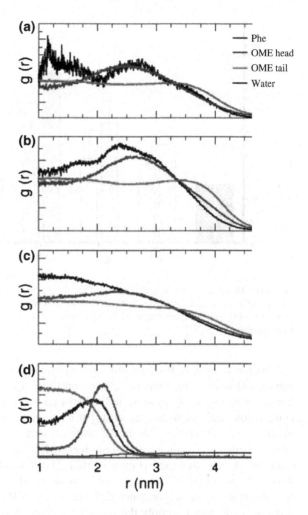

Fig. 5.10 Model of PAH-FAME nanoparticle located on *top* of a phospholipid bilayer membrane (lung cell wall)

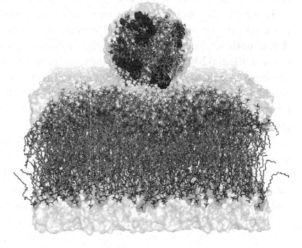

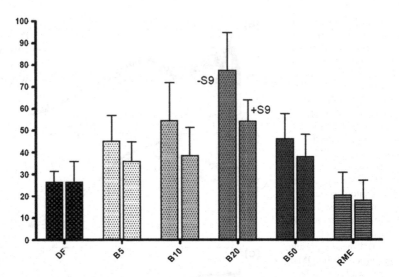

Fig. 5.11 Means and standard deviations of quadruple mutagenicity (Ames) tests of exhaust extracts of fossil diesel (DF), B5, B10, B20, B50, and pure biodiesel (RME), with (+S9) and without (−S9) metabolic activation of the test strain by rat liver enzyme. The figure is adapted from Munack et al. [33]

This mechanism of PAH-inclusion in nanoparticles and easier transport to, and subsequent entry into, lung cells, can constitute a possible explanation for the increased mutagenicity of collected exhaust particles from bio-blended diesel, in comparison with collected exhaust particles from the exhaust of pure fuels, observed as early as in 1994 in the Swedish study at MTC [11]. In that study, environmentally classified fossil diesel fuels (MK1) and (MK2) were blended with 5 and 30 % rape seed methyl ester (RME). The observed mutagenicity, using the Ames-test, was higher than expected from the results of the singular fuels. From this observation, the author concluded that blends of RME and MK1 react as new, different fuels, not as simply the sum of the components of the blend.

However, the unexpected consequence of increased mutagenicity from blending a fuel of renewable source into a fossil fuel, was not brought into the larger debate until Krahl et al. [24], more than a decade later, presented their results of the so-called B20-study. In that study, extracts of particles collected from the combustion exhaust of various blending ratios of biodiesel in fossil diesel were compared, in terms of mutagenicity measured with the Ames-test. All the blends provided mutagenicity results higher than the results for both pure fossil diesel and pure biodiesel. B20 expressed the highest mutagenicity (Fig. 5.11).

The increased mutagenicity of biodiesel blends, compared to pure biodiesel and pure fossil diesel, is not explained in Munack et al. [33], Krahl et al. [25] nor in Munack et al. [34]. However, the authors theorized that it might have something to do with aging of the biodiesel, i.e., that the biodiesel used in the experiments was old, and thus expressed an increased precipitation of certain oligomers, co-occuring with observed increased presence of the antioxidants carotenoids and tocopherols, in the blends. A likely mechanism for the increased mutagenicity was however not elucidated.

Recent findings also report that blending biodiesel into fossil diesel results in vehicle exhausts with increased cytotoxicity and interleukin (IL)-6 release [10]. This was observed when comparing exhausts from vehicles fuelled with either unblended fossil diesel (B0) or with 50 % biodiesel (B50). The results were obtained in transformed human bronchial epithelial (BEAS)-2B cells.

The observed results described above indicate how the MDS-supported theory of the formation of PAH-FAME nanoparticles might constitute a plausible mechanism for the phenomenon. Many PAHs are well-known mutagens, that can explain the results of the B20-study, while cytotoxic and IL-6 release are classical cellular immune responses to foreign substances, of which PAHs are examples of.

This new type of health damaging emission component (PAH-FAME nano-particle), which has been elucidated and modeled in Figs. 5.7 and 5.8 constitute an unintended consequence for health and the environment, from the implementation of a renewable energy. It presents a clear example of the damaging effects that can result from a failure to follow the precautionary principle in the implementation of alternative fuels. Furthermore, extensive use of the strategy of blending biofuels into regular fuels, without first elucidating the consequences, should therefore be avoided.

References

1. Andersen O (2003) Transport and industrial ecology: problems and prospects. Ph.D. thesis. Vestlandsforsking, Sogndal. Aalborg University, Aalborg. http://www.vestforsk.no/en/reports/transport-and-industrial-ecology-problems-and-prospects. Accessed 17 Sep 2013
2. Andersen O, Lundli H-E, Brendehaug E , Simonsen M (1998) Biodiesel in heavy-duty vehicles—Strategic plan and vehicle fleet experiments. Final report from European Commission ALTENER-project XVII/4.1030/Z/209/96/NOR. Sogndal: Western Norway Research Institute. http://www.vestforsk.no/filearchive/rapport-18-98.pdf. Accessed 17 Sep 2013
3. Andersen O, Manzetti S , Spool D van der (2012) Bio-blending of diesel might impact exhaust toxicity. In: The '7th International conference on the environmental effects of nanoparticles and nanomaterials', September 10-12. Poster. The Banff Centre, Banff. http://www.oens.ualberta.ca/wp-content/uploads/2011/12/7th-ICEENN-Abstracts-2012.pdf. Accessed 17 Sep 2013
4. Crutzen PJ, Ehhalt DH (1977) Effects of nitrogen fertilizers and combustion on the stratospheric ozone layer. Ambio 6(2/3):112–117 (Nitrogen: a special issue)
5. Dunn R, Bagby M (1994) Aggregation of unsaturated long-chain FattyAlcohols in nonaqueous systems. J Am Oil Chem Soc 72(1):123–130
6. European Commission (2010a) Directive 2009/28/EC of 23 Apr 2009 on the promotion of the use of energy from renewable sources and amending and subsequently repealing directives 2001/77/EC and 2003/30/EC. European Commission
7. European Commission (2010) EUROPE 2020 A strategy for smart, sustainable and inclusive growth. Communication From The Commission. Brussels, European Commission

8. Forster P, Ramaswamy V, Artaxo P, Berntsen T, Betts R, Fahey D, Haywood J, Lean J, Lowe D, Myhre G, Nganga J, Prinn R, Raga G, Schulz M , Van Dorland R (2007) Changes in atmospheric constituents and in radiative forcing. In: Climate change 2007: the physical science basis. Contribution of working group I to the fourth assessment report of the intergovernmental panel on climate change. Cambridge University Press, Cambridge. http://www.ipcc.ch/publications_and_data/ar4/wg1/en/ch2s2-10-2.html. Accessed 17 Sep 2013

9. Gasparatos A, Stromberg P (2012) Socioeconomic and environmental impacts of biofuels. Evidence from developing nations. Cambridge University Press: New York

10. Gerlofs-Nijland M, Totlandsdal A, Tzamkiozis T, Leseman D, Låg M, Schwarze P, Ntziachristos L and Cassee F (2013) Cell toxicity and oxidative potential of engine exhaust particles—impact of using particulate filter or biodiesel fuel blend. Environ Sci Technol http://pubs.acs.org/doi/pdf/10.1021/es305330y. Accessed 17 Sep 2013

11. Grägg K (1994) Effects of environmentally classified diesel fuels, RME, and blends of diesel fuels and RME on the exhaust emissions. Report MTC 9209. AB Svensk Bilprovning Motortestcenter, Haninge, sweden pp 1–44

12. Hess B, Kutzner C, van der Spool D, Lindahl E (2008) GROMACS 4.0: algorithms for highly efficient, load-balanced, and scalable molecular simulation. J Chem Theory Comput 4:435–447

13. IEA (2010) Sustainable production of second-generation biofuels: potential and perspectives in major economies and developing countries. International Energy Agency, Paris

14. Jorgensen WL, Chandrasekhar J, Madura J, Impey R, Klein M (1983) Comparison of simple potential functions for simulating liquid water. J Chem Phys 79:926–935

15. Wl Jorgensen, Maxwell D, Tirado-Rives J (1996) Development and testing of the OPLS all-atom force field on conformational energetics and properties of organic liquids. J Am Chem Soc 118:11225–11236

16. Kaminski G, Friesner R, Tirado-Rives J, Jorgensen W (2001) Evaluation and reparametrization of the OPLS-AA force field for proteins via comparison with accurate quantum chemical calculations on peptides. J Phys Chem B 105:6474–6487

17. Khanna M, Ando A, Taheripour F (2008) Welfare effects and unintended consequences of ethanol subsidies. Rev Agric Econ 30(3):411–421

18. Kinoo B (1996) Biodiesel demonstration in Belgium. Final report from ALTENER project XVII/4.1030/93-22. VITO-report. Vlaamse Instelling voor Technologisch Onderzoek. Belgium

19. Kleywegt G, Jones T (1998) Databases in protein crystallography. Acta Crystallogr Sect D-Biol Crystallogr 54:1119–1131

20. Knothe R, Bagby M (1995) 13C-NMR Spectroscopy of unsaturated long-chain compounds: an evaluation of carbon signals as rational functions. J Chem Soc Perkin Trans 2(3):615–620

21. Knothe R, Bagby M, Weisleder D, Peterson R (1994) Allylic mono- and di-hydroxylation of isolated double bonds with selenium dioxide-tert-butyl hydroperoxide. NMR characterization of long-chain enols, allylic and saturated 1,4-diols, and enones. J Chem Soc Perkin Trans 2:1661–1669

22. Knothe R, Dunn R and Bagby M (1994) Surface Tension Studies on Novel Allylic Mono- and Dihydroxy Fatty Compounds. A method to distinguish erythro/threo diastereomers. J Am Oil Chem Soc 72(1): 43–47

23. Kousoulidou M, Ntziachristos L, Fontaras G, Martini G, Dilara P, Samaras Z (2012) Impact of biodiesel application at various blending ratios on passenger cars of different fueling technologies. Fuel 98:88–97

24. Krahl J, Munack A, Ruschel Y, Schröder O, Bünger J (2008) Exhaust gas emissions and mutagenic effects of diesel fuel, biodiesel and biodiesel blends. In: Proceedings of the SAE Powertrains, Fuels, and Lubricants Meeting, 6–9 Oct 2008. Chicago, IL: SAE International. Chicago http://papers.sae.org/2008-01-2508/. Accessed 17 Sep 2013

25. Krahl J, Munack A, Schmidt L, Petchatnikov M, Schröder O (2010) Wechselwirkungen zwischen Biodiesel und modernen Dieselkraftstoffen. Paper presented at the 8. FAD-Konferenz Herausforderung—Abgasnachbehandlung für Dieselmotoren. Beiträge. Dresden: Förderkreis Abgasnachbehandlungstechnologien für Dieselmotoren e.V. pp 209–223 3, 4 Nov 2010

26. Landels R, Harold S, Gill R (1995) Fuel additives for biodiesel. In: Proceedings of international conference on standardisation and analysis of biodiesel, Vienna, 6–7 Nov 1995

27. Lee J, Garcia-Ulloa J, Koh L (2012) Biofuel expansion in Southeast Asia: Biodiversity impacts and policy guidelines. Socioeconomic and environmental impacts of biofuels. evidence from developing nations. Cambridge University Press, New York , pp 191–204

28. Lewis RJ (1996) Sax's dangerous properties of industrial materials, 9th edn. Van Nostrand Reinhold, New York

29. Manzetti S (2012) Ecotoxicity of polycyclic aromatic hydrocarbons, aromatic amines, and nitroarenes through molecular properties. Environ Chem Lett 10:349–361

30. Manzetti S, Andersen O , Czerwinski J (2011) Biodiesel, fossil diesel and their blends: chemical and toxicological properties. biodiesel: blends, properties and applications. Nova publishers, New York, pp 41–68. https://www.novapublishers.com/catalog/product_info.php?products_id=21023. Accessed 17 Sep 2013

31. McGrattan B (1994) Examining the decomposition of ethylene vinyl acetate co-polymers using TG/GC/IR. Appl Spectrosc 48(12):1472–1476

32. McKenzie C, Godwin A, Morawska L, Ristovski Z, Jayaratne E (2005) Effect of fuel composition and engine operating conditions on polycyclic aromatic hydrocarbon emissions from a fleet of heavy-duty diesel buses. Atmos Environ 39:7836–7848

33. Munack A, Krahl J, Bünger J, Ruschel Y , Scröder O (2008) Exhaust gas emissions and mutagenic effects of modern diesel fuels, GTL, biodiesel, and biodiesel blends. paper presented at the IGR — International conference of agricultural engineering XXXVII Congresso Brasileiro de Engenharia Agrícola Brazil, Aug 31 to Sept 4, 2008

34. Munack A, Krahl J, Schröder O, Bünger J (2010) Potentials of biofuels. Paper presented at the XVIIth world congress of the international commission of agricultural and biosystems engineering (CIGR) Québec City, 13–17 June

35. Oleksiak S, Stepien Z, Urzedowska W, Czerwinski J , Andersen O (2010) Influence of bio-components content in fuel on emission of diesel engines and engine oil detorioration. Euro Oil & Fuel 2010. Biocomponents in diesel fuels — impact on emission and aging on engine oil. Paper presented at the Euro Oil and Fuel 2010. Crakow: Oil and Gas Institute Krakow, pp 7–14

36. Rapier R (2008) Renewable diesel. Biofuels, solar and wind as renewable energy systems. Benefits and risks. Springer, 153–170

37. Ratcliff M, Dane A, Williams A, Ireland J, Luecke J, McCormick R, Voorhees K (2010) Diesel particle filter and fuel effects on heavy-duty diesel engine emissions. Environ Sci Technol 44:8343–8349

38. Rathbauer J (1995) Fatty Acid Methyl Ester (FAME) as special winter fuel. In: Chartier P, Beenackers A, Grassi G (eds) Biomass for energy, environment, agriculture and industry, Proceedings of the 8th European biomass conference, Elsevier. Vienna, 3-5 Oct 1995, pp 1174–1177

39. Ribeiro NM, Pinto AC, Quintella CM, da Rocha GO, Teixeira LSG, Guarieiro LLN, do Carmo Rangel M, Veloso MCC, Rezende MJC, da Cruz RS, de Oliveira AM, Torres EA and de Andrade JB (2007) The role of additives for diesel and diesel blended (ethanol or biodiesel) fuels: a review. Energy Fuels 21(4): 2433–2445

40. Schifter I, Díaz L, Rodríguez R (2010) Cold-start and chemical characterization of emissions from mobile sources in Mexico. Environ Technol 31(11):1241–1253

41. Schmidt C (2007) Biodiesel: cultivating alternative fuels. Environ Health Perspect 115(2):A86–A91

42. Schuettelkopf A, van Aalten D (2004) PRODRG - a tool for high-throughput crystallography of protein-ligand complexes. Acta Crystallogr D 60:1355–1363

43. Silvius M, Kaat A, Van de Bund H and Hooijer A (2006) Peatland degradation fuels climate change. An unrecognised and alarming source of greenhouse gases. Wetlands International, Wageningen
44. van der Spool D, Lindahl E, Hess B, Groenhof G, Mark A, Berendsen H (2005) GROMACS: fast, flexible and free. J Comput Chem 26:1701–1718
45. Tieleman D, Spool D van der and Berendsen H (2000) Molecular dynamics simulations of dodecylphosphocholine micelles at three different aggregate sizes: micellar structure and chain. J Phys Chem B 104 (27)l: 6380–6388
46. Turkcan A, Canakci M (2011) Combustion characteristics of an indirect injection (IDI) diesel engine fueled with ethanol/diesel and methanol/diesel blends at different injection timings. Paper presented at the sustainable transport, world renewable energy congress 2011. Linköping University Electronic Press, Linköping, Sweden, pp 3565–3672. http:// www.ep.liu.se/ecp/057/vol13/009/ecp57vol13_009.pdf. Accessed 17 Sep 2013
47. Wilson D (1997) Improving the quality of rapeseed methyl ester (RME) by the use of lubrizol's performance chemicals. Paper at the first international colloquium fuels, 16–17 Jan 1997, Ostfildem
48. Zehner O (2011) Unintended consequences of green technologies. In: Robbins P et al (eds) Green Technology, Sage. London, pp 427–432

Chapter 6
Towards the Use of Electric Cars

Abstract In this chapter, the historical development in the use of electric cars is described within a global content as occuring in five evolutions or waves, since 1835. The last two waves are described in more detail through the following case studies, the French VEL car and the Norwegian THINK car. They each had their own series of setbacks including. Problems in electro-chemistry, which have caused important limitations, as they have through the whole history of electrical cars. Unintended consequences in the form of increased human toxicity, freshwater eco-toxicity, freshwater eutrophication, and metal depletion, from potential increases in the future number of electric vehicles, are explained. Finally, a critical view on the policies for promoting the use of electric cars is presented.

6.1 The History of the Electric Vehicle in Five Waves

The phasing in of electric vehicles into the transport sector has proven a salient lesson, underscoring how prevailing patterns of consumption, in this case transportation habits, can slow down the implementation of alternative energy solutions. With regard to electric vehicles, a major limitation has been poor battery capacity, which reduce distance vehicles can cover before they need recharging. The global implementation of electric vehicles has not been straightforward, but unexpectedly, it has experienced five major waves of development [10].

The first wave of development began around 1835 and lasted until 1880. This was a period of many important inventions, and the first vehicle that was driven by an electric motor was demonstrated in the US, Scotland and the Netherlands in 1835. An important step was taken in 1859 with the introduction of the lead-acid battery, which still is the most common type of battery in cars today. Early in the 1870s, the first battery-powered electric cars came into regular use. It would take nearly two decades before the first vehicle on the road with an internal combustion engine was demonstrated by Carl Benz, in 1885 [19].

O. Andersen, *Unintended Consequences of Renewable Energy*,
Green Energy and Technology, DOI: 10.1007/978-1-4471-5532-4_6,
© Springer-Verlag London 2013

The second wave of development occurred during the period 1890–1925, and includes the so-called "golden period" (1895–1910) for electric vehicles [1]. This was a period with a relatively large number of electric cars on the road, just as many as the combustion engined vehicles. Around 1910, there were 30,000 electric cars in the US. The golden period faded with the launch of Ford's T-model, which made use of a combustion engine. During the first world war there was a resurgence of carbon battery electric vehicles because of oil rationing, particularly for civilian activities. At that time the USA became a major exporter of electric cars to Europe.

The turn of the last century also saw the development of the first hybrid cars with both electric and combustion engines. Several car manufacturers were involved, not the least the French, who exhibited a hybrid vehicle at an automobile and cycle show in Paris in 1901 [1]. France had, at this time, a well-developed system for charging electric vehicles, with a total of 265 charging stations operating in 1900. These took the form of coin-operated dispensers that delivered the required amount of electricity to charge the battery.

The third wave of development lasted from about 1935 to 1950 and was strongly connected with the second world war. As was also the case during the first world war, oil usage was prioritized for war purposes. However, a new development that came to be of great importance: namely the increased use of coal for domestic electricity production. It was especially the case in the UK where coal was used in distributed electricity networks. This was instrumental for the implementation of the world's most comprehensive fleet of electric vehicles, the so-called the Milk Float. Throughout Britain these electric milk vans were used for the door–to-door delivery of milk to households across the country (Fig. 6.1). After the II World War, the fleet consisted of about 70,000 electric vehicles in the UK. The Milk Float continued to be in operation over the next decades, while the use of other electric cars again faded.

Fig. 6.1 A view of an electric milk delivery vehicle of the Ludgershall Dairy float from 1950, a Douglas Electric vehicle, made at Kingswood in Bristol by the Douglas motorbike company. Thanks to Peter Davenport for the photo

The fourth wave took place in France, and pertains to the VEL-experiments, an entirely new concept which is described in more detail in Sect. 6.2 below. This wave was based on a new concept. In the years 1965–1980, there was a growing focus on reducing local air pollution within cities and the 1970s especially, saw an emerging awareness that new forms of energy based on renewable resources, were needed. All major automakers had programs for the development of electric vehicles to meet these needs. But the programs also had symbolic value: it had become important for automakers to demonstrate environmental responsibility. There were advances made in electrochemistry and battery technology, but as we will see in the VEL-experiments, the interest was to fizzle out once again.

The fifth wave of development lasted from about 1987 to 2000. In Norway, this wave was tightly connected to the emergence of the electric car manufacturer Think, described in Sect. 6.3 below. The focus was again on improving battery technology, and quite optimistic projections were launched. But the period was dominated by a new public debate. Although the environmental debate on the necessity for reductions in local air pollution was still important, there was a new problem that emerged with considerable force: man-made climate change. This was also a period with new challenges connected to concepts such as sustainable development, sustainable energy systems and sustainable mobility. To face these challenges, new policies were developed, and international agreements made. This was a period of tighter legislation, both resulting from global climate conventions and regional regulations, such as the Zero Emission Vehicles regulations in California [9].

6.2 Electric Vehicle Case 1: VEL-experiments in France

To improve the understanding of unintended consequences of the implementation of alternative fuels it is useful to look more closely at the large-scale introduction of electric vehicles in France, which was attempted in the so-called "Vehicule ELectrique" (VEL) experiments. Early in the 1970s, a group of engineers at Electricité de France (EDF) launched ambitious scenarios for a transition to the use of electric cars instead of vehicles with internal combustion engines. These scenarios included an ambitious vision for the development and use of hydrogen-powered fuel cell vehicles. The implementation of this experiment has previously been described and analyzed by Callon [3, 4].

The EDF engineers were inspired by the student revolts in 1968, which started in France, as well as environmental issues such as air pollution and noise from road traffic. The scenario assumed that people would change their values, habits, and expectations of the cars, and that they would work together to achieve the common goal of an environmentally friendly electrified transport system.

The engineers redefined the production system for cars, where EDF would be responsible for the development of the electric engines, while Renault were marginalized to only producing the chassis and car body. This was a hard blow at one of Europe's major industrial companies and also toward the close link between the private car and the petroleum industry.

The VEL-experiments in reality never left the laboratories and drawing boards. Various problems soon emerged, which Hughes [12] described as "reverse salients":

> Reverse salients are components in the system that have fallen behind or are out of phase with the others. (...) Reverse salients are comparable to other concepts used in describing those components in an expanding system that is in need of attention, such as drag, limits to potential, emergent friction, and systemic inefficiency [12 p. 73].

Reverse salients are thus a notation for the particularly cumbersome challenges and limitations in certain segments of the system that function as barriers to the whole system's development and expansion. These relate to what Callon [4] terms "guerilla warfare." The first reverse salients in the VEL-experiments appeared in connection with the basic system components. No established system existed for the handling of used batteries. The development of well-functioning fuel cells also turned out to be quite problematic, the new expensive catalytic materials had a tendency to quickly become contaminated. Engineers at Renault presented counter expertise, by pointing out the necessity to build up a complete new infrastructure, with service stations for the handling of used catalytic material for fuel cells, and for the recycling of used batteries. They questioned who would be responsible for this.

The Renault engineers went further and pointed out that during the 3 years since the start of the VEL-experiments, the social protest movements had slowed down, at least those who were opposing the private car based society. Plans for the reindustrialization of the automotive industry was instead laid out, with more environmentally friendly buses and low-emission cars. Callon emphasizes in his analysis that the conventional automobile industry was fully rehabilitated due to the initiatives of the Renault engineers, failing fuel cell development and a fading protest movement. It was now the VEL researchers who were silenced, as their efforts completely lost credibility and strength.

6.3 Electric Vehicle Case 2: Think, the Norwegian Electric Car

The Norwegian electric car manufacturer Think provides an interesting case study that illustrates the ups and downs of electricity as an energy carrier for passenger cars [10].

The end of car production at Think can be considered a significant setback for the introduction of electric vehicles. Despite the fact that Norway had not previously had any significant automotive production earlier, Think had become one of the most common electric vehicles available on the global market during the 1990s. This was the outcome of a long history of innovation. During the 1950s and

1960s technologies to form molded plastic, especially for recreational boats, became well established. The innovator Lars Ringdal at the company Bakelitt-fabrikken AS had, by 1973, created two electric vehicle prototypes, utilizing molding technology. He did not refer to them as cars, rather as vehicles with four wheels, steering, and an electric engine [10].

However, it was not until the end of the 1980s, with a new focus on energy and environmental issues that the story continues. The backdrop was a growing discourse on sustainable development and in particular sustainable energy. The environmental problems caused by energy use in motorized transport, especially in cities, were constantly being debated. It was in 1988, just a year after the launch of the concept of sustainable development in the Brundtland Commission report "Our Common Future" [18], that the Norwegian authorities considered it appropriate to take presenting a leading step, on the international stage, and follow through on the recommendations in the report. A preparatory conference for the Rio 1992 Earth Summit was already planned in Bergen for 1990. Representatives from the Norwegian authorities recalled the 1973 prototypes and Ringdal was contacted to discuss the possibility of a Norwegian electric car being put into production. The idea was supported by Ringdal and not the least by his equally innovative son, Jan Otto Ringdal. The planning started and support was secured from RCN.

The first prototype of the new generation electric car was produced in 1991. It was highly innovative, with a design that combined aluminium and molded plastic technology. The basic idea was that both the engine technology and the plastic materials in the body would be environmentally friendly. All materials would be easy to recycle. The next step was to produce a car that could be possible to use under normal driving conditions. The 1994 winter Olympic Games in Lillehammer was considered a good international showcase, especially since it had already been launched as the first environmentally friendly Olympic Games. During the Games there were severe restrictions on all polluting private transportation in the town of Lillehammer. This made electric cars an interesting option, and the agreement to produce 12 cars was finalized.

In the town of Lillehammer, with steep slopes and temperatures down to minus 20 °C, the cars worked fine, but with the lead-acid batteries, only with the more expensive nickel–cadmium batteries. The cars had thus been very expensive to produce, with a cost of at least 60,000 € each.

Showcasing at the Lillehammer Olympics worked well for generating interest in electric vehicles. Prince Albert of Monaco on attending the games, took one of the cars home and showed it off during the Monaco Grand Prix. A first contact with interested parties in USA was established. After the introduction of strict emission standards in California, the USA was seen as a strong potential market. The California Air Resources Board (CARB) had passed the Zero-emissions vehicle (ZEV) mandate in 1990, which required the seven major automobile suppliers in the United States to offer electric vehicles in order to continue sales of their gasoline powered vehicles in California. Nearly 5,000 electric cars were designed and manufactured by GM, Toyota, Honda, Ford, Nissan, and Chrysler, which then, failing to pass the safety standards, was later destroyed or donated to museums and educational institutions.

In the period 1995–1996, approx. 140 new electric vehicles were produced by Think. The car was awarded several prizes at a rally for electric vehicles between Gothenburg and Oslo. Most of the Think cars were used by Norwegian companies, but 45 of them were taken to California. They did however end up with the same ill-fated destiny as the American electric vehicles, not fulfilling the safety criteria, they where only used as test vehicles for a period of 2 years.

The next generation electric vehicles from Think—produced for major markets—was ready in 1998. The Think car had always been a small car, with only two seats and almost no luggage space. It had also been marketed only as a niche car for city driving. This was a clear limitation. The advantage was that it was produced as an electric vehicle from the start. This resulted in a car that worked better than most other types of electric vehicles. Electric vehicles manufactured by the major automakers were mostly ordinary cars equipped with electric motor and adapted batteries, which proved problematic. A battery with a weight of 200–300 kg cannot easily fit into a normal car. However, there were other limitations as well. Originally, the Think vehicle was intended to have a range of 100 Km between charges. But due to the added weight of the batteries, it appeared that this was difficult to achieve. It had to be reduced to 85 km, which at best would be 70 km under real driving conditions with hilly roads, heavy winter, cold weather, and extra electricity for heating of the car's passenger compartment.

While Think struggled financially, the USA car maker Ford showed interest in the electric vehicles, first at an exhibition in Brussels in 1998. A Think car was taken to the Ford headquarters in Detroit. The top leadership of Ford was excited about the car and Ford became the major shareholder in 1999, taking full ownership the following year. Optimism remained high and a new factory was opened by Norway's king and prime minister.

With the American take-over, problems emerged. The problems turned out to be serious reverse salients, and came as "guerrilla attacks" from within the Ford Company. The company argued that American car buyers did not like the dull plastic surface. They were accustomed to glossy cars. They also had much larger equipment requirements than common in Europe. The Think car was initially designed to be a very simple car. American cars had, however, dual air bags, power steering, ABS brakes, air-conditioning, and electrical windows. Compliance with this would mean fundamental changes, not only in the production, but for the vehicle concept itself. The Think cars were in addition equipped with nickel–cadmium batteries. Americans did not want cadmium, which is a toxic metal. Also the nickel–cadmium batteries needed to be removed and undergo extensive maintenance/recharging after every 6,000–7,000 km of driving, less in most cases before less than 1 years operation. Americans were not used to such a high level of maintenance, and were only familiar with having lead-acid batteries in their automobiles. The expertise at Ford was also limited to this type of battery.

It was thus necessary to implement a series of compromises in order to produce a new generation of Think cars. The most serious of these was the use of lead-acid batteries, and their associated poor performance. The expected battery life of 3 years was, under real driving conditions, no more than 1–1.5 years. This was not at all acceptable, and thus the development of a new type of battery was initiated. But it was not be long before new problems emerged. In 2002, the implementation of the strict emission and energy requirements in California were repeatedly being postponed. The USA market for electric cars was not as promising as it looked before. Battery development did not proceed as expected, and the USA auto industry began to lose interest in electric vehicles. The focus was now on fuel cell technology and hydrogen. Ford decided to discontinue its Think department in 2003. After several changes in ownerships, including the Finnish vehicle company Valmet, the Norwegian investment group InSpire, and the Spanish company Electric Mobility Solutions, production was terminated in March 2011, and the company filed for bankruptcy on June 22, 2011 [2].

The production and uptake of electrical vehicles has a more favorable climate in Norway. These vehicles are exempt from value added tax, road tolls, and they can be used in priority lanes (car-pooling, bus, and taxi lanes). This is very convenient for their owners who save both time and money. However, private individuals in Norway who own an electric vehicle are generally quite affluent, residing in Oslo's high income west side. It is not their only car, rather the second or third in the household. The drivers are mostly alone in the car, travelling to and from work, and get the most from avoiding the time consuming traffic jams drivers of regular cars have to face.

6.4 Some General Observations From the Two Case Studies

The development of improved batteries for electric vehicles has been much slower than expected. The defining characteristic of the electrical vehicle industry has been a clear over-optimism and technological oversell.

The two case studies show the need to analyze electric vehicle implementation as a complex and heterogeneous whole, from the start. The concept of technological systems by Hughes [11, 12] proves to be useful in this context. We are dealing here with three different technological systems: 1) the transport system, 2) the car production system, and 3) the energy production system. These three systems are connected, but they can also be considered as separate technological systems.

The concept of reverse salient is helpful in facilitating a more nuanced understanding of the importance of various barriers arising during the implementation of alternative energy in general, and electric vehicles in particular.

6.5 Toxicity and Metal Depletion

With a massive increase to be expected in the future number of electric vehicles, there are concerns that the production of the vehicles will lead to serious unexpected consequences, in terms of increased human toxicity, freshwater eco-toxicity, freshwater eutrophication, and metal depletion [7]. In particular, the production of electronic equipment necessary for an electrical vehicle requires a variety of metals, which poses a challenge for recycling and raises serious toxicity concerns [14]. The life cycle human toxicity potential of electrical cars has been estimated to be 180–290 % higher than for cars with internal combustion engines [7]. This stems, to a large degree, from the high use of copper wires in electrical vehicles, and the use of nickel in cars with lithium-nickel cobalt-manganese batteries. The toxic emissions connected with the excavation of these metals occur, for the greater part, during the disposal of sulfide mine tailings. Similar elevated life cycle results from electric vehicles has been found for their ecotoxicity potential in freshwater systems.

Depletion of scarce metals is also a commonly sited concern with electric vehicles (see for example Gaines and Nelson [5, 6]). The life cycle metal depletion potential for electrical vehicles has been found to be 2–3 times higher than for combustion-based vehicles [7].

6.6 The Policies for Promoting Electric Cars: Are they Working as Intended?

Holtsmark [8] questions if the electrical car policies give the intended results. He claims that there is a widespread but naive notion that sufficiently large subsidies for alternative energy, whether it is wind energy, biofuels or electric cars, leads to reduced use of fossil energy. He points out that it is not that simple. This is due the fact that subsidy policy, which largely dominates environmental policies, particularly in Norway, is ineffective and has several unintended effects. In many cases the effects are directly opposite to their purpose. However, the policies are still being implemented, because the willingness for action is so high that there is a tendency to produce counterproductive solutions rather than no solutions at all [17].

Many people seem to be convinced that electric cars will take over from petrol and diesel cars. Currently, it is only in Norway that electric cars constitute more than just a marginal share of total car sales, about 2 % in 2011 [8]. In the rest of the world, electric vehicle sales are on the per mill level. But this could change quickly, as it is possible that countries like China and USA, in their efforts to become less dependent on imported oil, will implement measures aiming to increase the sales of electric vehicles. These are, however, countries where coal power constitutes a large share of the electricity grid mixture, occasionally resulting in higher life cycle CO_2 emissions for electric cars than for petrol cars

[13, 15]. It cannot be expected that the CCS technology will be implemented on a large scale in the coming decades, so coal combustion with large CO_2 emissions is likely to remain the main production technology for electricity in many countries [8]. Currently, there are no trends suggesting that more electric cars will provide significant, if at all any, reductions in life cycle CO_2 emissions. It is thus highly dubious that the electric car represents an environmentally friendly solution in the long run. Not only is the electricity grid mixture in large parts of the world dominated by electricity generated from coal, but a major technological development in relation to batteries are needed. Today the weight of a battery, such as in a Nissan Leaf, is about 300 kg, giving the car a range of only 110 km [8].

The most concern with current policy instruments aimed at electric vehicles is that this gives households the incentive for obtaining a second or third car. Should the number of households with two or three cars increase substantially, increased environmental impacts along many lines will be the result.

The many environmental and societal impacts of transport, including pollution, accidents, congestion, and seizure of valuable land, can, according to Holtsmark [8], only be solved by making it more costly to use the transport system, not by making it cheaper to buy and use electric vehicles. Electric car owners should have to pay for use of roads, parking lots and the energy they use, as has also been concluded by the Norwegian Ministry of Finance [16 p. 8]. It is therefore very difficult to justify why electric vehicles have continued access to priority lanes (car-pooling, bus, and taxi lanes) in Norway and elsewhere.

References

1. Anderson CD, Anderson J (2005) Electric and hybrid cars: a history. McFarland & Company. http://books.google.no/books/about/
Electric_and_Hybrid_Cars.html?id=nMMRzDe1YcoC&redir_esc=y. Accessed 17 Sep 2013
2. Bolduc D (2011) Norwegian EV maker Think files for bankruptcy. Automotive News. http://www.autonews.com/apps/pbcs.dll/article?AID=/20110622/COPY01/306229798/
1193#axzz2VQRUbV4J. Accessed 06 Jun 2013
3. Callon M (1980) The state and technical innovation: a case study of the electrical vehicle in France. Res Policy 9(4):358–376
4. Callon M (1987) Society in the making: the study of technology as a tool for sociological analysis. The social construction of technological systems: new directions in the sociology and history of technology. MIT Press, London, pp 83–103
5. Gaines L, Nelson P (2009) Lithium-ion batteries: possible materials issues. Argonne National Laboratory, Argonne
6. Gaines L, Nelson P (2010) Lithium-ion batteries: examining material demand and recycling issues. Argonne National Laboratory, Argonne
7. Hawkins T, Singh B, Majeau-Bettez B, Strømman A (2013) Comparative environmental life cycle assessment of conventional and electrical vehicles. Int J Ind Ecol 17(1):53–64
8. Holtsmark B (2012) Elbilpolitikken – virker den etter hensikten? (The electric vehicle policy – is it working as intended?). Samfunnsøkonomien 5:4–11
9. Høyer KG (1999) Sustainable mobility-the concept and its implications. PhD thesis, Department of Environment, technology and social studies, Roskilde University, Roskilde

10. Høyer KG (2007) The battle of batteries: a history of innovation in alternative energy cars. Int J Altern Propul 1(4):369–384
11. Hughes TP (1983) Networks of power: electrification in western society, 1880–1930. John Hopkins University Press, Baltimore
12. Hughes TP (1987) The evolution of large technological systems. The social construction of technological systems: new directions in the sociology and history of technology. MIT Press, London, pp 51–82
13. Ji S, Cherry CR, Bechle MJ, Wu Y, Marshall JD (2012) Electric vehicles in China: emissions and health impacts. Environ Sci Technol 46(4):2018–2024
14. Johnson J, Harper E, Lifset R, Graedel T (2007) Dining at the periodic table: metals concentrations as they relate to recycling. Environ Sci Technol 41(5):1759–1765
15. Moyer M (2010) The dirty truth about plug-in hybrids. Sci Am 303(1):54–55
16. Norwegian Ministry of Finance (2007) En vurdering av særavgiftene (An assessment of excise taxes). NOU 2007:8. Departementenes servicesenter Informasjonsforvaltning
17. Risan L (2012) Slag og slagsider i klimaforskningen. (Battles and biases in the climate research). Nytt Norsk Tidsskrift 12:29–38
18. WCED (World Commission for Environment and Development) (1987) Our common future. Tiden Norsk Forlag, Oslo
19. Westbrook MH (2001) The electric car: development and future of battery, hybrid and fuel-cell cars. Institution of Electrical Engineers. http://books.google.no/books?id=6EfS 0jRFQTkC&pg=PA6&hl=no&source=gbs_toc_r&cad=4#v=onepage&q&f=false. Accessed 17 Sep 2013

Chapter 7
Solar Cell Production

Abstract Producing electricity by harvesting solar energy in photovoltaic (PV) solar cells can, as we will see, lead to serious unintended consequences. It is the manufacturing of the PV that causes the most evident impacts on health and the environment. The large consumption of water for rinsing wafers between etching steps, creates a subsequent need for wastewater treatment facilities. Production of the cells also leads to additional emissions of fluorinated compounds, such as hexafluoroethane (C_2F_6), nitrogen trifluoride (NF_3), and sulfur hexafluoride (SF_6). These are all extremely strong greenhouse gases, with global warming potentials (GWPs) of 9,200, 17,200, and 39,800 times that of CO_2. These emissions from PV manufacturing are worth serious consideration, as they are counter to the prevailing idea that solar PV cells are a very "green" energy technology. In fact, massive production of new PV solar panels will imply huge global emissions of greenhouse gases. Alternative, lower GHG emitting production processes are being developed, but they rely on the use of very toxic and explosive gases, such as fluorine (F_2).

7.1 Unexpected Consequences from Photovoltaic Cell Production

Solar photovoltaic cells (PV) are widely admired for the way they silently, and seemingly without pollution, extract clean energy from the sun's rays. However, this is by no means the full picture. There exist serious environmental issues connected not only to the way PV solar cell panels are currently manufactured, but also to the way they are disposed of at the end of their use phase [45].

The manufacturing of PV cells relies on the use of a large number of different toxic chemicals and explosive compounds [35, 18]. Therefore, PV cell production in general, has potential hazardous impacts from the chemicals being released during the various process steps, like volatile organic substances during panel laminating [35]. In addition, accidents connected to the operation of the factories,

O. Andersen, *Unintended Consequences of Renewable Energy*,
Green Energy and Technology, DOI: 10.1007/978-1-4471-5532-4_7,
© Springer-Verlag London 2013

Table 7.1 Life cycle external costs of electricity generation in Germany, expressed in € cent per kWh [11]

	Coal	Lignite	Natural gas	Nuclear	PV	Wind	Hydro
Noise	0	0	0	0	0	0.0050	0
Health	0.7300	0.9900	0.3400	0.1700	0.4500	0.0720	0.0510
Material	0.0150	0.0200	0.0070	0.0020	0.0120	0.0020	0.0010
Crops	0	0	0	0	0	0.0007	0.0002
Total	0.7450	1.0100	0.3470	0.1720	0.4620	0.0797	0.0522

can lead to severe health consequences for workers in the manufacturing facilities, as well as for local residents. The PV manufacturing industry is in addition one of the leading emitters of fluorinated compounds. This includes hexafluoroethane (C_2F_6), nitrogen trifluoride (NF_3), and sulfur hexafluoride (SF_6). These are all extremely strong greenhouse gases, with global warming potentials (GWPs) of 9,200, 17,200, and 39,800 times that of CO_2 on a 100 year time horizon [12].

The use of SF_6 is, for example, commonly used in order to remove damages to the wafers, so-called saw damages, that have occurred during the slicing of the crystalline silicon ingots.[1] SF_6 is also used during the oxide etching process [36]. The use of these strong greenhouse gases implies that there going to be a unintended impact on the climate system from the large scale manufacture of PV solar cells [3, 4, 14, 15]. This is in addition to the fact that production of silicon for crystalline PV solar cells also contributes substantially to the emission of GHGs [33]. These aspects explain why electricity production with the use of PV solar cells is worse than many other technologies, in terms of life cycle GHG emissions, as shown in Fig. 1.1. Actually, the average value of 39 g CO_{2eq}/kWh for solar PV in Fig. 1.1 might be on the low side, as even more detrimental life cycle performance have been reported, with emissions as high as 110 g CO_{2eq}/kWh [15, 33, 17]. On top of this, the European Commission ExternE project on external costs for energy presented life cycle emissions from PV installations in Germany to be as high as 180 g CO_{2eq}/kWh [11, 13]. In ExternE use the use of PV solar cells was ranked even worse than natural gas in terms of life cycle GHG emissions and air pollution [12, 14, 15]. As shown in Table 7.1, of the studied German electricity generation options, only lignite and coal was worse regarding the combined external costs from impacts related to noise, health, material, and crops in the life cycle of the technologies. The two other renewable energy technologies wind and hydro, came out substantially better [11].

In addition to the impacts described above, some of the materials entering into PV cells are of concern regarding the unintended consequences from their increased usage. One example are fullerenes, such as C_{60}, also termed

---

[1] The ingot is a casted chunk of crystalline silicon that is made from a molten smelt. It is cut into thin slices that ends up as wafers after several steps of processing.

Buckminster fullerene[2] or in short Bucky balls. These carbon nanomaterials are used in certain new types of solar cells [25, 26]. There have been a series of studies into toxicological properties and adverse health aspects of fullerenes, with regard to the way they interfere with cellular mechanisms, both in humans and animals [16, 30, 31, 44]. However, much remains to be known about the impact of these carbon molecules on the human body and the environment.

A different, but also very serious environmental impact of State of the Art (SoA) PV cell production, is its high water consumption. The wet chemical process, which is SoA in PV cell production, involves the processing of the wafers in a series of consecutive but different, chemical baths. Between each chemical treatment, the wafers need to be rinsed free from chemicals, using tens of thousands of liters of clean water in the process. In 2011, a leading USA solar company with manufacturing plants in Malaysia and the Philippines stated that their wet chemical-based process required about 15,000 liters of high purity process water per minute for a 1.4 GW production facility for PV solar cells [9].

As such, major chemical waste treatment facilities are necessary to cope with the large volumes of contaminated water from the PV solar cell production when using the wet chemical process [2]. That is why alternative process types, using less water, are needed. This is the topic of the next sub-chapter.

7.2 Alternative Production Processes

As an alternative to the current wet chemical etching process used in crystalline PV solar cell production, dry plasma-based processes are being developed [36, 1, 8, 22–24, 34, 37]. Some of these processes use fluorine (F_2), which is very toxic, and actually characterized as a poison gas [21]. This yellow gas is extremely reactive and a very powerful caustic irritant to human skin. According to the EC Scientific Committee on Occupational Exposure Limit Values (SCOEL) it is due to its strong reactivity with hydrogen-containing compounds, that fire and explosion might result from exposure to the gas [39]. It is thus a very dangerous fire and explosion hazard. Emissions can only be avoided through stringent control measures, which include on-site generation, isolation of the process from human exposure, and air scrubbers to remove the fluorine from escaping into the environment [38, 40].

There is a potential that emissions of fluoride-containing compounds from the production plants are having an impact on human health and the environment [32]. The use of wet scrubbers is a means to prevent gases escaping into the ambient air during the process. Many chemical entities can however potentially be present in

---

[2] The name Buckminster fullerene is homage to the American designer, author, and inventor Richard Buckminster "Bucky" Fuller, as C_{60} resembles the geodesic domes that were his trademark.

the discharge water from the scrubbers. They include various products from the reaction between fluorine-containing compounds and water.

The toxicological profiles of the fluoride-containing gases, which characterizes their toxicological and adverse health effects, have been compiled by U.S. Department of Health and Human Services, Public Health Service at the Agency for Toxic Substances and Disease Registry [41]. In addition to the GHG-gases there are three types of fluorine-containing emission that need to be considered concern. These three are also connected to alternative PV production processes, such as dry chemical wafer etching, and will be described in the next sub-chapters. They are

- Hydrogen fluoride (HF)
- Fluorine (F_2)
- Silicon fluorides (SiF_x)

7.3 Hydrogen Fluoride (HF)

Hydrogen fluoride, when inhaled, is poisonous to humans [21]. It is a corrosive irritant to skin, eyes, and mucous membranes. It also has teratogenic and reproductive effects, also give rise to skeletal damage. Hydrogen fluoride (HF) gas dissolves in water, where it ionizes into fluoride (F^-). The discharge of water from wet scrubbers is thus a source of F^- emissions into waterways.

7.4 Fluorine (F_2)

Fluorine is characterized as a poison gas [21]. It is a very powerful caustic irritant to tissue. It is also a very dangerous fire and explosion hazard. In a wet scrubber there are several possible reactions between fluorine gas and water, as shown below:

(I) $F_2(g) + H_2O(l) \rightarrow HOF(g) + HF(g) [T = -40\,°C]$

(II) $2\ HOF\ (g) \rightarrow 2\ HF(g) + O_2(g)\ [T = +20\,°C]$

(III) $F_2(g) + H_2O\ (l) \rightarrow 2\ HF(g) + \frac{1}{2}O_2(g)\ [T = +20\,°C]$
$\quad$ + small amounts of ozone

(IV) $F_2(g) + 2\ H_2O\ (l) \rightarrow 2\ HF\ (g) + H_2O_2 [pH < 7]$

(V) $2\ F_2(g) + 2\ OH^- \rightarrow 2\ F^- + OF_2(g) + H_2O \rightarrow O_2(g) +$
$\quad 4\ F^- + 2\ H^+ [pH\ approximately\ 14]$

Reaction I shows the formation of the chemical compound hypofluorous acid (HOF). HOF is an intermediate in the oxidation of water by F_2 and decomposes at room temperature to HF and O_2 as shown in reaction II and the resulting netto reaction III. The hydrogen peroxide (H_2O_2) formed in reaction IV decomposes exothermically in the scrubber effluent water and oxygen according to reaction VI.

(VI) $2H_2O_2 \rightarrow 2\ H_2O$ (l) $+\ O_2(g)$

The scrubber must have enough water through-flow to avoid the concentration of hydrogen peroxide reaching levels where an explosion might occur. In very alkaline conditions the strongly oxidizing compound oxygen difluoride (OF_2) is formed, as shown in reaction equation V.

7.5 Silicon Fluorides (SiF_x)

Silicon tetrafluoride (SiF_4) is the most significant gas emission, of the three gases HF, F_2, SiF_4 and constitutes more than 90 % of the total Emissions from dry PV etching. Other silicon fluoride than SiF_4 (here given the common name SiFx) can be present in the emissions, however, these compounds constitute only minor components only minor components of the emission gas mixture.

In the air SiF_4 is formed when hydrogen fluoride reacts with silica. When present in the air SiF_4 is characterized as a poison and is moderately toxic by inhalation [20]. Furthermore, it is a corrosive irritant to skin, eyes, and mucous membranes.

In the wet scrubbers the emissions of SiF_4 can react with water in the same manner as it can react with moisture in the air, to form silicon oxide (SiO_2), HF [42], and fluorosilicic acid (H_2SiF_6), according to Tylenda [40] and [21]. This is shown in reaction equations VII and VIII.

(VII) $SiF_4 + 2\ H_2O \rightarrow SiO_2 + 4\ HF$
(VIII) $3SiF_4 + 2\ H_2O \rightarrow SiO_2 + 2\ H_2SiF_6$

HF, after its ionizing, contributes to the release of F^- into the environment through the scrubber effluent water. Therefore, the environmental impact of F^- in water is a consequence of SiF_4 releases.

In the USA and Canada H_2SiF_6 is commonly used in drinking water fluoridation. This has spurred a debate on health concerns. The basis for this was a series of reports on the relationship between water treated with H_2SiF_6 and elevated levels of lead (Pb) in children's blood [7, 27, 28]. This association has been attributed to both the effect that incompletely dissociated fluorosilicates in drinking water have on increasing the body's cellular uptake of Pb [27, 28] and to increased corrosion of Pb-bearing plumbing by fluorosilicates [7]. However, the scientific rigor of these studies and the validity of their conclusions have been questioned by Urbansky and Schock [42]. They argue that fluorosilicates are hydrolyzed before reaching the consumer's tap and therefore have no effect on the solubility and bioavailability of lead. The safety of using H_2SiF_6 in water fluoridation is however not established. Studies of the chronic effects of H_2SiF_6 use in water fluoridation are conducted. Safety studies on fluoride have only been conducted using pharmaceutical-grade sodium fluoride, not industrial-grade silicon fluorides [6]. Thus, the use of H_2SiF_6 harvested from wet scrubbers, could potentially result in the indicated unintended consequences IX.

A third reaction between silicon tetrafluoride and water is shown in the reaction equation IX.:

(IX) $SiF_4 + 4 H_2O \rightarrow H_4SiO_4 + 4 HF$

The orthosilicic acid (H_4SiO_4) being formed is stable at pH < 3.2. At less acidic conditions it undergoes successive conversion as shown in equation X.:

(X) $H_4SiO_4 \rightarrow H_{2n+2}Si_nO_{3n+1} \rightarrow (H_2SiO_3)_n \rightarrow (H_6Si_4O_{11})_n$
     $\rightarrow (H_2Si_2O_5)_n \rightarrow (SiO_2)_n$

Regulations for fluorides in the urine of workers exist in the USA, where ACGIH has an established biological exposure index (BEI) with values of 3 and 10 mg/g creatinine prior to and at the end of shift respectively [5.].

7.6 Positive Unintended Consequences of Solar Energy

After all these negative consequences, it is worth noting that as well as providing renewable energy, PV solar cell technology might also encourage changes in household energy consumption. Research has shown that there is a 'double-dividend' effect of using PV solar systems in UK households [19]. The installation of PV systems for harvesting solar energy encouraged households to reduce their overall electricity consumption by approximately 6 %. In addition, a shift in demand to times of peak solar PV electricity generation was observed.

References

1. Agostinelli G, Choulat P, Dekkers HFW, De Wolf S, Beaucarne G (2005) Advanced dry processes for crystalline silicon solar cells. paper presented at the photovoltaic specialists conference, 2005. Conference record of the thirty-first IEEE. IEEE, pp 1149–1152. http://b-dig.iie.org.mx/BibDig/P05-0850/pdffiles/ papers/283_625.pdf. Accessed 17 Sep 2013
2. Agostinelli G, Dekkers HFW, De Wolf S and Beaucarne G (2004) Dry Etching and Texturing Processes for Crystalline Silicon Solar cells: Sustainability for Mass Production. paper presented at the 19th European photovaltaic solar energy conference. Paris, 423–426. http:// www.docin.com/p-51912614.html. Accessed 17 Sep 2013
3. Alsema EA (2000) Energy pay-back time and CO2 emissions of PV systems. Prog Photovoltaics Res Appl 8(1):17–25

4. Alsema EA and de Wild-Scholten M (2005) Environmental impact of crystalline silicon photovoltaic module production. Paper presented at the materials research society symposium, G: life cycle analysis tools for" green" materials and process selection. Boston

5. American Conference of Governmental Industrial Hygienists (ACGIH) (2000) Documentation of the threshold limit values and biological indices. American Conference of Governmental Industrial Hygienists, Cincinnati

6. Connett P (2012) 50 Reasons to oppose fluoridation. fluoridealert.org - Fluoride Action Network September. http://www.fluoridealert.org/articles/50-reasons/ Accessed 28 June 2013

7. Coplan M, Patch S, Masters R, Bachmann M (2007) Confirmation of and explanation for elevated blood lead and other disorders in children exposed to water disinfection and fluoridation chemicals. Neurotoxicology 28(5):1032–1042

8. Dresler B, Köhler D, Mäder G, Kaskel S, Beyer E, Clochard L, Duffy E, Hoffman M , Rentch J (2012) Novel Industrial single sided dry etching and texturing process for silicon solar cell improvement. Poster presented at the 27th European photovoltaic solar energy conference and exhibition. Frankfurt, Germany

9. Duffy E (2012) SOLNOWAT - Development of a competitive zero global warming potential (GWP) dry process to reduce the dramatic water consumption in the ever expanding solar cell manufacturing industry. D7.3 exploitation plan (interim). www.SOLNOWAT.com. Accessed 17 sep 2013

10. European Commission (1995) ExternE: externalities of energy. Prepared by ETSU and IER for DGXII: science, research and development, Study EUR 16520-5 EN, Luxembourg

11. European Commission (2003) External costs. research results on socio-environmental damages due to electricity and transport. European Commission, Brussels (Directorate-General for Research Information and Communication Unit)

12. Forster P, Ramaswamy V, Artaxo P, Berntsen T, Betts R, Fahey D, Haywood J, Lean J, Lowe D, Myhre G, Nganga J, Prinn R, Raga G, Schulz M, Van Dorland R (2007) Changes in atmospheric constituents and in radiative forcing. In: Climate change 2007: the physical science basis. contribution of working group i to the fourth assessment report of the intergovernmental panel on climate change. Cambridge University Press, Cambridge. http://www.ipcc.ch/publications_and_data/ar4/wg1/en/ch2s2-10-2.html. Accessed 17 Sep 2013

13. Fthenakis VM, Kim HC (2007) Greenhouse-gas emissions from solar electric- and nuclear power: A life-cycle study. Energy Policy 35:2549–2557

14. Fthenakis VM, Kim HC (2010) Photovoltaics: life-cycle analyses. Sol Energy 85(8):1609–1628

15. Fthenakis VM, Kim HC, Alsema E (2008) Emissions from photovoltaic life cycles. Environ Sci Technol 42:2168–2174

16. Fujita K, Morimoto Y, Ogami A, Myojyo T, Tanaka I, Shimada M, Wang W, Endoh S, Uchida K, Nakazato T, Yamamoto K, Fukui H, Horie M, Yoshida Y, Iwahashi H, Nakanishi J (2009) Gene expression profiles in rat lung after inhalation exposure to C_{60} fullerene particles. Toxicology 258:47–55

17. Jungbluth N (2005) Life cycle assessment of crystalline photovoltaics in the Swiss ecoinvent database. Prog Photovoltaics Res Appl 13:429–446

18. Kuemmel B, Krüger Nielsen S, Sørensen B (1997) Life-cycle analysis of energy systems. Roskilde University Press, Roskilde

19. Kairstead L (2007) Behavioural responses to photovoltaic systems in the UK domestic sector. Energy Policy 35:4128–4141

20. Lewis RJ (1996) Sax's dangerous properties of industrial materials, 9th edn. Van Nostrand Reinhold, New York

21. Lewis RJ (ed) (1997) Hawley's Condensed Chemical Dictionary. 13th ed. John Wiley & Sons, Inc

22. Linaschke D, Leistner M, Mäder G, Grählert W, Dani I, Kaskel S, Lopez E, Hopfe V, Kirschmann M, Frenck J (2008) Plasma enhanced chemical etching at atmospheric pressure for crystalline silicon wafer processing and process control by in-line FTIR gas spetroscopy. EU PVSEC Proceedings. Paper presented at the 23rd European photovoltaic solar energy

conference and exhibition. Valencia, Spain: WIP Wirtschaft und Infrastruktur GmbH and Co Planungs KG, Sylvensteinstr. 2, 81369 München, Germany, 1907 – 1910

23. Lopez E, Beese H, Mäder G, Dani I, Hopfe V, Heintze M, Moeller R, Wanka H, Kirschmann M, Frenck J (2007) New developments in plasma enhanced chemical etching at atmospheric pressure for crystalline silicon wafer processing. The compiled state-of-the-art of PV solar technology and deployment. In: Proceedings of the international conference. Paper presented at the 22nd European photovoltaic solar energy conference and exhibition (EU PVSEC). European Commission, Joint Research Centre, Milan

24. Lopez E, Dani I, Hopfe V, Wanka H, Heintze M, Möller R, Hauser A (2006) Plasma enhanced chemical etching at atmospheric pressure for silicon wafer processing. In: Poortmans J (ed) 21st European photovoltaic solar energy conference 2006, European Commission, Joint Research Centre -JRC. Proceedings of the international conference held in Dresden (CD-ROM), Germany, 4–8 Sept 2006. WIP-Renewable Energies, Dresden, 1161–1166. http://publica.fraunhofer.de/documents/N-63736.html. Accessed 28 June 2013)

25. Manzetti S, Andersen O (2012) Toxicological aspects of nanomaterials used in energy harvesting consumer electronics. Renew Sustain Energy Rev 16(1):2102–2110

26. Manzetti S, Behzadi H, Andersen O, van der Spool D (2013) Fullerenes toxicity and electronic properties. Environ Chem Lett 11(2):105–118

27. Masters R, Coplan M (1999) Water treatment with silico-fluorides and lead toxicity. Int J Environ Sci (China) 56:435–449

28. Masters R, Coplan M, Hone B, Dykes J (2000) Association of silicofluoride treated water with elevated blood lead. Neurotoxicology 21(6):1091–1100

29. Nakagawa Y, Suzuki T, Ishii H, Nakae D, Ogata A (2011) Cytotoxic effects of hydroxylated fullerenes on isolated rat hepatocytes via mitochondrial dysfunction. Arch Toxicol 85:1429–1440

30. Oberdörster E, Zhu S, Blickley T, McClellan-Green P, Haasch M (2006) Ecotoxicology of carbon-based engineered nanoparticles: effects of fullerene (C_{60}) on aquatic organisms. Carbon 44:1112–1120

31. Ozsvath D (2009) Fluorides and environmental health: a review. Rev Environ Sci Biotechnol 8:59–79

32. Pehnt M (2006) Dynamic life cycle assessment (LCA) of renewable energy technologies. Renewable Energy 31:55–71

33. Photovoltaics World (2011) Champions of photovoltaics: cells and modules. Photovoltaics World (October)

34. Phylipsen G, Alsema EA (1995) Environmental life-cycle assessment of multicrystalline silicon solar cell modules. Padualaan 14, NL-3584 CH Utrecht, Utrecht University, The Netherlands

35. Piechulla P, Seiffe J, Hoffman M, Rentsch J and Preu R (2011) Increased ion energies for texturing in a high-throughput plasma tool. In: Proceedings at the 26th European photovoltaic energy conference. paper presented at the 26th European photovoltaic energy conference. Hamburg

36. Rentsch J, Jaus J, Roth K, Preu R (2005) Economical and ecological aspects of plasma processing for industrial solar cell fabrication. In: IEEE Xplore. Digital library. Conference record of the thirty-first IEEE. IEEE, pp 931–934

37. Schottler M, de Wild-Scholten M, Ruess S (2010) Gas abatement for crystalline silicon solar cell production. Photovoltaics Int (PV-Tech) 9:20–28

38. SCOEL (1998) Recommendation from scientific committee on occupational exposure limits for fluorine, hydrogen fluoride and inorganic fluorides (not uranium hexafluoride). European Commission

39. Stockman P (2009) Going green with on-site generated fluorine: sustainable cleaning agent for PECVD processes. In: Proceeding of: Photon's 4th production equipment conference Munich March 2009.

40. Tylenda CA (2003) Toxicological profile for fluorides, hydrogen fluoride, and fluorine (Update). U.S. Department of Health and Human Services. Public Health Service. Agency for Toxic Substances and Disease Registry, Atlanta
41. Urbansky E, Schock M (2000) Can fluoride affect lead (ii) in potable water? Hexafluorosilicate and fluoride equilibria in aqueous solution. Int J Environ Stu 57:597–637
42. Voltaix (2006) Material safety data sheet for: silicon tetrafluoride (SiF4). Voltaix, LLC Post Office Box 5357, North Branch, New Jersey 08876-5357. http://www.voltaix.com/images/doc/Mssi030_Silicon_Tetrafluoride.pdf. Accessed 28 June 2013
43. Yamago S, Tokuyama H, Nakamura E, Kikuchi K, Kananishi S, Sueki K, Nakahara H, Enomoto S, Ambe F (1995) In vivo biological behavior of a water-miscible fullerene: [14]C labeling, absorption, distribution, excretion and acute toxicity. Chem Biol 2:385–389
44. Zehner O (2011) Unintended consequences of green technologies. In: Robbins P et al (eds) Green technology. Sage, London, pp 427–432

Chapter 8
Final Discussion and Conclusions

8.1 Reduction and Proximity

As has been shown, the development of renewable energy is connected to a large number of negative, unintended consequences. However, it is critical to also consider the *reduction of energy use*, as a key issue in this discussion, irrespective of whether the energy is taken from renewable or nonrenewable sources. This was touched upon in Chap. 2, which illustrated the unintended consequences of energy-efficiency measures and the rebound effects of other actions aiming to reduce energy consumption.

The potential for improving the efficiency of energy conversion, in an effort to reduce the total energy use, is large. It has, for example, been assessed that through all the energy conversion processes in the USA, more of the total energy is lost than what is used [1]. 58 exajoules was lost, while 45 exajoules was in the form of useful energy in 2005.

However, positive consequences of energy-efficiency improvements have also been experienced, one such example is the shift to LED municipal lighting as part of a suit of energy-efficiency measures employed in cities. An unintended consequence of this measure was reduced maintenance and thus less disturbance to traffic flows, because the LEDs failed less frequently than the old bulbs they replaced [2].

Physical proximity is another important factor for saving energy. In older cities, it was a common practice to build dwellings shoulder-to-shoulder, facilitating energy transfer and utilization between accommodation levels. Such close proximity had the unintended consequence of bringing people closer together in new ways, creating walkable neighborhoods that promoted social interaction.

The proximity principle applied to energy production and use was mentioned in Chap. 1. Reduction of transportation by providing goods and services as close as possible to the consumers, is needed to attain a society based on sustainable consumption. This was already agreed upon by the Commission on Sustainable Development meeting in Oslo in January 1994 [3]. Reducing transportation is a key issue to reduce energy consumption.

O. Andersen, *Unintended Consequences of Renewable Energy*,
Green Energy and Technology, DOI: 10.1007/978-1-4471-5532-4_8,
© Springer-Verlag London 2013

It is interesting to note that people who voluntarily reduce their material and energy consumption substantially, often unexpectedly discover new interests and experience higher satisfaction with their low consumption lifestyles [2].

8.2 Peak Phosphate

Just as the concept of *peak oil* is connected to the excavation of fossil oil resources, *peak phosphate* has relevance for renewable energy sources. Phosphate rock is the source of phosphorous, which is essential for the growth of plants. It is used by plants to form the cell walls and membranes through structural components known as phospholipids. Phosphate groups are in addition key parts of the backbone of the DNA molecule. Production of artificial fertilizers consumes phosphate, which is incorporated into the fertilizer. Production of bioenergy is dependent on the use artificial fertilizers. It has been estimated that by 2025–2030 the world will reach peak phosphate, with a flattening out of fertilizer production.

It is commonly understood that phosphate rock, like oil and other key nonrenewable resources will follow a Hubbert production curve. A key difference between peak oil and peak phosphorus is however that oil can be replaced with other forms of energy once it becomes too scarce. However, there is no substitute for phosphorus in food production [4]. This will have a catastrophic impact, for example, without phosphate US corn production will be cut in half [5]. Thus, food production could flatten out, which could create severe conditions for a growing global population.

An aggravating fact of the expected phosphate shortage is that the global presence of phosphate rock is highly concentrated in the Western Sahara, which has been an occupied territory, since 1976. When Spanish colonialists left Morocco invaded two-thirds of the country, and the last third in 1979 after Mauritania withdrew. Morocco now controls more than 85 % of all high-grade phosphate in highly disputed land. Many neighboring states reject the Moroccan administration of Western Sahara, and several states have established diplomatic relations to the "Sahrawi Arab Democratic Republic" represented by the Polisario Front. This movement is operating in exile in Algeria, and UN recognizes it as the rightful representative of the territory. It is believed that the phosphate deposits were the major reason that Morocco took an interest in the Western Sahara. The Polisario Front would like to have it back [5].

8.3 Critique of the Concept of Unintended Consequences

Some theorists argue that due to the many negative, unintended consequences of governmental policies, governmental intervention should be limited [2]. Others claim that the application of the concept of unintended consequences is politically motivated and suspect. If legislators suspend an activity in order to avoid the unintended consequences, then the positive impacts of that activity are also going to be lost.

There is also an argument that all human activities yield unanticipated consequences and that strong governance, even when imperfect, is necessary to prevent more series harm to the environment.

8.4 Reflexive Thinking About Renewable Energy

Reflexive thinking based in the concept of the 'risk society' [6], can be as much about exploring possibilities as avoiding danger and unintended negative impacts of technology. Renewable energy solutions are becoming more important as fossil energy sources are depleted. Therefore, the technologies for harvesting renewable energy must be improved upon, so that they do not degrade the environment and resource basis for the future.

There are also distributional societal aspects pertaining to the development of renewable energy technologies. Since few, if any, renewable energy technologies come without some negative impact, it appears pertinent to ask how the impacts fall across society [7]. Regarding the benefits of new technology, it is just as relevant to ask how the distribution of the benefits looks like. It is argued that typically, the people who get the most negative impacts from a technology, are not the principal recipients of the benefits of that technology. Jeffery Thomas Morris [7] looks at this in relation to nanotechnology, but a parallel can be drawn to renewable technologies. So as a final note, the unfair correlation between recipients of impacts and benefits is also an unintended consequence of new technologies connected to renewable energy.

References

1. Beck U (1992) Risk society: towards a new modernity. Sage, Newbury Park, CA
2. Cordell D, Drangert J-O, White S (2009) The story of phosphorus: global food security and food for thought. Glob Environ Change J 19(2):292–305
3. Høyer KG (1997) Recycling: Issues and Possibilities. In: Brune D, Chapman D, Gwynne M and Pacyna J (eds) *The Global Environment. Science, Technology and Management.* Weinheim: VCH Verlagsgesellschaft mbH, 817–832
4. Morris JT (2010) Risk, language, and power: the nanotechnology environmental policy case. PhD thesis, Falls Church, VA, Virginia Polytechnic Institute and State University. http://scholar.lib.vt.edu/theses/available/etd-10042010-225927/unrestricted/Morris_JT_D_2010_v2.pdf

5. Pearce F (2011) *Phosphate: A Crit Res Misused and Now Running Low*. New Haven, CT: Yale University. http://e360.yale.edu/feature/phosphate_a_critical_resource_misused_and_now_running_out/2423/. Accessed 28 June 2013
6. Whitesides G, Crabtree G (2007) Don't forget long-term fundamental research in energy. Science 315:796–798
7. Zehner O (2011) Unintended consequences of green technologies. In: Robbins P et al (eds) Green Technology, Sage, London, pp 427–432

Printed in the United States
By Bookmasters